HISTORY, PHILOSOPHY AND SOCIOLOGY OF SCIENCE

Classics, Staples and Precursors

HISTORY, PHILOSOPHY AND SOCIOLOGY OF SCIENCE

Classics, Staples and Precursors

Selected By

**YEHUDA ELKANA
ROBERT K. MERTON
ARNOLD THACKRAY
HARRIET ZUCKERMAN**

ENDLESS HORIZONS

By Vannevar Bush

ARNO PRESS

A New York Times Company

New York — 1975

Reprint Edition 1975 by Arno Press Inc.

Reprinted from a copy in
    The Newark Public Library

HISTORY, PHILOSOPHY AND SOCIOLOGY OF SCIENCE:
Classics, Staples and Precursors
ISBN for complete set: 0-405-06575-2
See last pages of this volume for titles.

Manufactured in the United States of America

———◆———

Library of Congress Cataloging in Publication Data

Bush, Vannevar, 1890-
    Endless horizons.

    (History, philosophy and sociology of science)
    Reprint of the ed. published by Public Affairs Press,
Washington.
    Includes index.
    1. Science. 2. Technology. I. Title. II. Se-
ries.
Q172.B87 1975   500   74-26253
ISBN 0-405-06581-7

ENDLESS HORIZONS

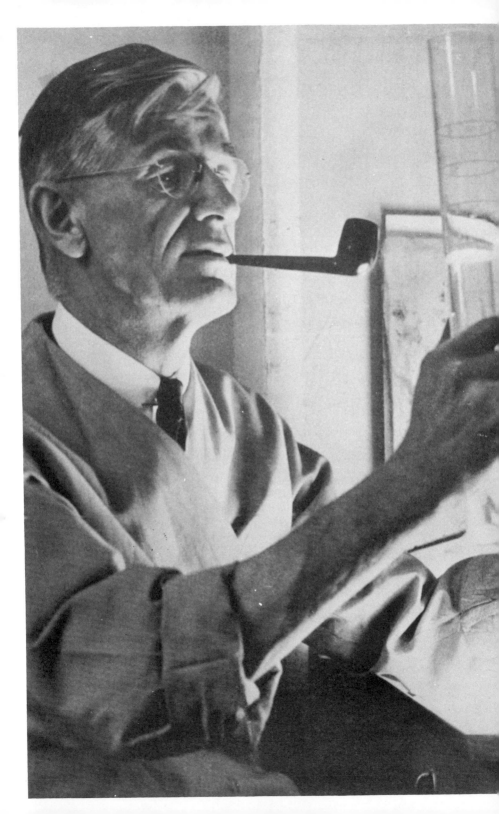

ENDLESS HORIZONS

By Vannevar Bush

Introduction By Dr. Frank B. Jewett

Public Affairs Press

WASHINGTON, D. C.

Public Affairs Press, 2153 Florida Ave., Washington, D. C. Copyright, 1946, by American Council on Public Affairs, M. B. Schnapper, Executive Secretary and Editor

INTRODUCTION

Dr. Vannevar Bush has served his country and the world with great distinction. During World War II he was the initial Chairman of the National Defense Research Committee and, since 1941, he has been the Director of the Office of Scientific Research and Development. As head of the latter agency he directed the mobilization of the entire civilian scientific and technical power of the nation and welded it together into the military establishment in the greatest industrial research and development organization man has ever known.

At the age of 55, Dr. Bush has achieved distinction in many fields of science and a position of power and esteem throughout the world which few if any men of science have ever attained to at any age.

What is especially striking about this extraordinary man is that he has over the years contributed significantly to our scientific knowledge and progress in many diverse ways—as a great engineer, as the creator of the most complicated and powerful mathematical tool ever devised, as the inventor of many ingenious devices, as the successful administrator of a jealously guarded democratic society of scholars, as the brilliant director of one of the world's greatest research organizations, and, finally, as the organizer, guiding spirit and driving force of the nation's scientific endeavors in a global war.

Clearly here is an unusual man, one in whom nature has seen fit to compound harmoniously many elements, not the least of which are those human characteristics which command the respect, admiration and loyalty of men who are themselves able, respected and admired. It is these intensely human characteristics which implement intellectual ability and give men endowed with them power to accomplish great things in great times like the one the world is passing through.

In attempting in a limited space to voice an appraisal of Dr.

Bush, I do not know what I can do better than quote some passages from a statement I made several years ago in presenting him for the Edison Medal of the American Institute of Electrical Engineers:

"So varied are the fields to which he has turned the attention of his fertile mind, and so uniformly has that attention served to enrich them, that a poll among those who know him and his work would develop a startling diversity of opinion as to first place in the order of his claims to distinction. One has but to go through the titles of his published papers and addresses and the long list of his patents to be impressed by the catholicity and power of his mind. Those of us who know him well, however, know that the published record represents but a small fraction of his diversified interests. There is little or nothing in this published record of his equipment for automatically regulating the proper sprinkling of plants in a greenhouse; of his experiments in photography and with viviparous tropical fish; of his prowess as a navigator and skipper of ocean cruisers, or in the scientific commercial raising of turkeys. These and many others, I suspect, he looks upon as relaxation avocations.

"If you are interested in forming an appraisal from the permanent record you must supplement the list of his scientific and technical work—largely in the field of electricity and engineering—by his contributions to technical education and its administration. You must read his reports on the patent system, his testimony before Congressional committees, and above all, his annual reports as president of the Carnegie Institution of Washington.

"It is a far cry from the ability required to design apparatus for transferring heat mechanically, or a lubricating system for vehicle wheel bearings, to that required to conceive and design the most powerful mathematical engine ever devised. By all customary standards it is a farther cry still from the type of mind which can devise a new electronic tube of a complicated radio set to one which can administer and direct creatively a great diversified fundamental science undertaking like the Carnegie Institution. Beyond that even is the greater gap which separates the quiet of the laboratory from the turmoil of the directorship of the Office of Scientific Research and Development of a nation at war, where all that is potential in civilian science and technology has to be forged for a maximum of conflict power in a minimum of time and with a minimum of conflict in the forging, where the job to be done is only partially one of dealing with inanimate things and largely one of dealing with strong and frequently perverse men. I never cease to marvel at a man who in days filled with such turmoil can find the time, and, above all, the philosophical detachment needed to formulate a state paper like the recent letter to Senator Kilgore on the problem of technological mobilization.

"Even if time permitted, I am not competent to appraise and judge most of Dr. Bush's scientific contributions, except insofar as they can be judged by their end results. Rather, I find it pleasanter to deal with the man himself as I know him.

"I first came to know Vannevar Bush at New London, Connecticut, during World War I, when, as a young man, he came to the antisubmarine laboratory. Even then I was struck with those characteristics which I was to find later were the hallmarks of the man—a brilliant versatile mind; an intense interest in and curiosity about everything that came within his ken; an uncanny ability to illuminate the dark corners of any matter to which he gave his attention; his fairness, simplicity and lack of ostentation; and above all, his complete intellectual honesty.

"While I know he wasn't born there, I always think of Bush as a Cape Codder. His roots are deep in the soil of that historic bit of sterile land on which our forefathers landed, and, metaphorically speaking, his shoes are full of its sand. Like thousands of others before him who have gone a-pioneering on land and sea, he returns there periodically to replenish that sand and so insure keeping himself a realist in a too frequently unrealistic world.

"Better than most men he can fight vigorously with any of the weapons men employ and do it without rancor. I have yet to see a case in which there was any aftermath of hard feeling or in which the feeling of respect and affection for him was not enhanced in the erstwhile antagonist.

"What I have sought to convey is the impression of a very able, very human man in whom are combined great ability, great energy, a great sense of reality, and the power of absolute truth, together with all the common qualities of men, including an ability to make mistakes. Unlike many men, he can and does admit error without losing face either with himself or anyone else; rather, he enhances it."

The details of Dr. Bush's life and some of his achievements are listed briefly in *Who's Who in America* and *American Men of Science*. The essential facts are these: Dr. Bush was born in Everett, Massachusetts, March 11, 1890. He was graduated from Tufts College in 1913 with the degree of B.S. and M.S. and in 1916 he received the degree of Eng. D. simultaneously from Harvard and Massachusetts Institute of Technology. He joined the faculty of M.I.T. in 1919, became Professor of Electrical Engineering in 1923, and Vice President and Dean of Engineering in 1932. In 1939 he was elected President of the Carnegie Institution of Washington.

The papers here assembled give but a general outline of the range of Dr. Bush's interests. However, they show something of his basic philosophy and of his methods of approach—the one well thought out and consistent; the other, independent, direct and incisive.

That much of his forward looking picture is destined to produce controversy is inevitable and in some cases, I suspect, intended. Of one thing I am certain, however, and that is that however much men may differ from him they will not question his motives nor his intellectual honesty.

Some day someone who can do justice to it will write the story of Dr. Bush's achievements as the Director of OSRD and of the debt the nation owes him.

FRANK B. JEWETT,
President, National Academy of Sciences.

ACKNOWLEDGMENTS

Brought together at the suggestion of the American Council on Public Affairs, the material in this book is drawn chiefly from the author's recent writings, speeches, and reports with regard to the problems and opportunities confronting science. Portions of the original texts have to some extent been revised and adapted in terms of those considerations which are of especial importance at the present time.

The sources of the various chapters are cited below:

"The Inscrutable Past," *Technology Review*, January, 1933.

"As We May Think," *Atlantic Monthly*, July, 1945.

"A Program for Tomorrow," "The War Against Disease," "The Public Welfare," "Renewal of Scientific Talent," "Reconversion Opportunities," "The Means to the End," *Report to the President*, July, 1945.

"Research on Military Problems," statement before the Select Committee on Postwar Military Policy, House of Representatives, January 26, 1945.

"Control of Atomic Energy," statement before the Senate Special Committee on Atomic Energy, December 3, 1945.

"Research and the War Effort," address before the American Institute of Electrical Engineers, January 26, 1943.

"The Teamwork of Technicians," address before the National Institute of Social Sciences, May 23, 1945.

"The Qualities of a Profession," address before the American Engineering Council, January 13, 1939.

"Our Tradition of Opportunity," address before the American Institute of Electrical Engineers, January 26, 1944.

"The Need for Patent Reforms," Report of the Science Advisory Board's Committee on the Relation of the Patent System to the Stimulation of New Industries, April 1, 1935. (The other signers of the report, to whom grateful acknowledgment is made, were W. H. Carrier, Chairman of the Board of the

Carrier Engineer Corp.; D. M. Compton, industrial consultant; Dr. Frank B. Jewett, Vice President of the American Telephone and Telegraph Co., and President of the Bell Telephone Laboratories; and H. A. Poillon, President of the Research Corporation.)

"Science for World Service," address before the New York Herald-Tribune Forum on Current Problems, October 31, 1945.

Throughout the preparation of this book for publication, Mr. F. G. Fassett, Jr., Director of the Office of Publications, Carnegie Institution of Washington, provided invaluable counsel and assistance. Helpful advice was also extended by Mr. Carroll L. Wilson, Executive Assistant, and Dr. Lyman Chalkley, Assistant to the Director, Office of Scientific Research and Development.

CONTENTS

1: THE INSCRUTABLE PAST

A review of the mode of living of our forefathers, if it is to be useful, should be sympathetic in its attitude. The lapse of time often obscures the difficulties surrounding a former generation, and we are apt to smile at crudities when a just estimate should rather leave us to marvel that so much was accomplished with so little.

It is especially pertinent that we should review the technical accomplishments of another period only in the light of the contemporary science. Otherwise, we may well be guilty of a patronizing complacency, and as a result lose the benefit to be derived from a really analytical view of history.

Take the early Nineteen-Thirties as an example. From this distance the mechanical aspects of that time certainly appear grotesque; but, when we realize that this was the period when physics was in the throes of conflicting and essentially independent theories, the fact that applications were made at all is remarkable.

We read of the trials of the men of that day and wonder that they could have been apparently content with their mode of life, its discomforts, and its annoyances. Instead, we should admire them for having made the best of a hard situation, and treasure the rugged qualities which they exemplified. It is possible that by taking our minds back, divesting them of their modern knowledge, and then studying these bygone days in an attempt really to appreciate their true worth, we should lose some of our satisfaction with respect to the technical accomplishments of our own generation, and be better prepared for advance. At least it is worth the attempt.

1

Those were interesting times when the second Roosevelt was elected, and the world was in the midst of the last great economic readjustment. It was a time of transition, evident enough as we now regard it, but perhaps not wholly appreciated at the moment. It was marked by great extremes; the United States were just emerging from the prohibition experiment, and international affairs were chaotic. The system of distribution had nearly broken down, and there was little real control of production. No one really understood the monetary system under which the civilized world then tried to operate, and which was based on the curious process of laboriously digging gold out of one hole in the ground in order equally laboriously to bury it in another. Hence, it is illuminating to review the life of a plain citizen of the period and the nature of his environment.

THE HARDSHIPS OF DAILY EXISTENCE

Consider, for example, a professor in some northern urban university, and let us attempt to appreciate the sort of life he led, with a sympathetic attempt to evaluate the extent to which his efforts were circumscribed by the hardships and discomforts of his daily existence.

It is necessary at the outset to realize that he was in the peculiar position of being regarded by many of his fellow countrymen as of outstanding intellect; while at the same time his scale of living was decidedly middle-class. Yet he had much of comfort, in the way that comfort was then regarded, and undoubtedly he considered himself well off.

He probably owned an automobile, for example, and in this he proceeded from his home to his work. In many ways his car was embryonic, for it was, of course, a relatively new development. To get it under way, he first started the engine turning over while entirely free from any operative connection with the vehicle, and then he had to go through 14 separate motions with his hands and feet before getting the car up to full speed. It seems hard to realize that this situation could have been long tolerated, but actually it persisted for years. Moreover, these motions, of various pedals and a hand lever, had to be carried through in a rigid sequence and with a fairly careful timing of the operations. A clumsiness, such as performing one act out of proper sequence, would spoil the whole affair, and the professor would practically have to start all over. Yet people of all degrees learned this ritual and drove cars everywhere. Nor were these motor cars of the Nineteen-Thirties toys, for in a curious delusion that weight and riding qualities were inseparable, even the better grades of automobiles were built so heavy that they weighed several hundred pounds per passenger, a total weight of as much as 4,000 pounds being not unusual.

On nearly all highways traffic moved in both directions at the same time! Moreover pedestrians crossed these roads at the same level. There was no drying of streets in the cities, so that they were often wet, and, in extreme weather, covered with ice. Car speeds seldom got above 60 miles per hour, but under conditions which then obtained, there was a carnage and literally tens of thousands were killed yearly in the country. Right in the hearts of cities there were grade intersections of important streets and practically no elevated arteries. The law officer who was often stationed at points where important arteries intersected on a level, and who attempted to regulate traffic by whistling and waving his arms, was often a diverting spectacle. His antics are still recalled with amusement by some of my elderly colleagues. No wonder our professor arrived at work with his nerves somewhat frayed, and that the scientific writings of the time reflect a sort of general nervousness and a haste to publish fragmentary findings.

It is hard to suppress untoward amusement and to preserve the sympathetic attitude when one considers the clothing which our professor wore. It was put on in layers, and, while the lower layer was periodically washed, the other layers were often worn practically continuously until they disintegrated, with only infrequent dips into various solvents. In spite of this intentional protection of the top layers from contact with water, they sagged and stretched, and this tendency was ineffectually combated by occasional pressing by a highly heated metal implement. If caught out in the rain—for accidental sprinkling accelerated this process of deformation—the professor would don still another layer called a raincoat, so designed that it drained the water principally into his shoes. These shoes, by the way, were never thoroughly washed or even cleansed with solvents. They were daubed over occasionally with an impermanent varnish, which was given some specular reflection by rubbing with a cloth, but which was not really waterproof. The shoes were of natural leather, close-fitting and entirely unventilated, and fastened in position with lacing cords, which frequently became entangled. The fastening of his clothes generally was by buttons. Even though sleeves did not open, buttons were still retained in position at his wrists, where they were actually quite inconvenient, as a sort of atrophied appendage. He wore a collar about his neck, kept meticulously white by frequent changing, as a sort of obvious presentation of one thoroughly clean portion of clothing. Some of his collars were rendered stiff and irritating by saturation with vegetable starches. They were surrounded by highly colored ornamented bands which he tied in intricate knots and adjusted with careful precision.

The lenses by which his vision was corrected were wired to his ears, or else held on by pincers which gripped his nose. As his accommodation was faulty he carried two sets of lenses which he alternated in position on his face, carrying the spare in a little metal box. Some of his colleagues wore both sets of lenses at the same time, made into a combination called bifocals, so that they could produce the effect of changing lenses by tipping the head and inclining the eyeballs. Of course with this arrangement they went about with their feet and the ground in their vicinity in a perpetual haze.

In his office the professor ordinarily found other conditions hardly conducive to logical thought. Right in the midst of his most careful musings, anyone, not merely his chosen friends and colleagues, but literally anyone, could interrupt instantly by calling him on the telephone. A bell would ring in his office and convention demanded that he should immediately cease everything else and answer. There was no provision whatever by which a conversation ensued only when both parties had indicated willingness; even tradesmen in the city could initiate the ordeal. In answering, it was necessary that the professor practically wrap himself up in an instrument. He would hold one gadget to his ear and another to his mouth, and entangled in the connecting wires, proceed to try to talk. Some forms of equipment, then just going out, but strangely enough persisting longest in this country, even required the use of both hands as well as the vocal organs.

The sounds heard over the telephone of the day were recognizable, but hardly natural, for only a fifth of the useful frequency spectrum, or even less, was transmitted. In fact, people habitually listened to radio music in which there was less than one-third the spectrum present, although scant enthusiasm for the result has come down to us, except in the amusing and rhapsodic advertising of the times.

Our professor was bound to be fairly uncomfortable for other reasons. His office was heated in winter, of course, but in summer was left to its own devices with the window open to admit the dirt and noise of the city and the hot, humid air. Even in winter, while there might possibly have been some control of air temperature aside from his own chance manual regulation, there certainly was none of humidity. Also the walls of his office, with none too certain thermal insulation from the outside conditions, took up almost any temperature whatever, and the conditions for his bodily radiation varied in a wide and erratic manner. It is strange that these matters were so completely ignored, for they are hardly mentioned in the technical literature of the time, although certainly engineers should have been somewhat conversant with the laws of radiation, if not of biophysics.

Somewhere on the premises of the college there was a heating plant, which probably consisted of simply a coal fire with
some distribution medium such as steam. For every thermal
unit released by burning coal, there was transmitted into the
building actually less than a thermal unit, so that the process
was highly inefficient. Thermal pumps were in the early stages
of development, and the central stations played little part in
the heating of a city except perhaps to operate coal-burning
heating plants of their own and sell steam.

The most striking feature of our professor's day, however,
and one which he considered an important part of his work,
was the giving of lectures. He would stand up in front of a
group of 50 students or so and orally recite a bit of scientific
matter, perhaps meanwhile drawing crude diagrams on a large
flat black surface with a white crayon. This was done not alone
in the presentation of new thoughts and researches, but as a
means of imparting well-known information to students. It
would not disturb him in the least that a hundred other professors in various parts of the country might be doing exactly
the same thing at the same time. It was the custom of the day
that he should appear personally for this ceremony, although it
would have been possible even then to prepare a much more
finished presentation by vocal cinema, realizing of course that it
would not have been stereoscopic and that the articulation
might have been a bit crude. Thus a fair fraction of our professor's time was occupied in a rehash of the well-known before
large groups, where anything approaching Socratic dialogue
was patently impossible. Another amusing feature of these
lectures was the so-called taking of notes by the students, who
attempted by simple pencil and paper to reproduce important
ideas as the affair proceeded. The results were naturally fragmentary, but we must remember that under the then existing
system a word once spoken was lost unless recorded in some
such manner.

At noon the professor might take a walk, either because of
assumed benefit to the functioning of his organic processes, or

more likely as a temporary means of escape from the distractions of his office. If the latter incentive was his reason, the attempt was likely to be unsuccessful. The streets were heavy with the odor and smoke which nearby factories poured freely into the air almost without restraint. The price paid by the public for thus distributing unburned fuel on the breeze, and later abstracting it from draperies and clothing, was startling. One statistically-inclined person computed that no less than 70,000,000 tons of soot fell on the country every year, so that all the buildings were soon dingy and the sun was at times obscured.

He would also meet with din and confusion. Electric cars on rails, with hard steel wheels and steel spur gears, were still used for urban traffic. They made so much noise that one could literally be heard a mile away on a still night. Every automobile carried noise-making apparatus, usually in the form of a diaphragm operated by an electric vibrator, ostensibly for the purpose of cautioning pedestrians, although one wonders how an incautious one could have been extant. Automobile engines, burning highly volatile refined hydrocarbons, filled the air with carbon monoxide due to faulty combustion. Worse yet, there was a furor at about this time for the use of admixtures of anti-detonants. These were really useful in view of the nature of the fuel on the market commercially and its complete admission before compression, and a very popular type incidentally consisted of tetraethyl lead. Apparently these exhaust gases did little real harm to the pedestrians, although the data on the point were then incomplete, but the knowledge of their presence in the air was disturbing, at least to the scientifically minded.

An airplane or two may have roared overhead with unmuffled exhausts during this noon-hour walk, for there were quite a few planes in use. The professor probably watched these with a bit of trepidation. In the event of even minor power-plant difficulty they were obliged to land immediately and precipitately at high ground speed. There were no landing spots on buildings, nor could they have been used if present, for the planes of the commercial routes came to the ground while still traveling as much as 60 miles an hour. The slightest fog was a serious matter, for it could not be dissipated, and there was no way in which the pilot of the usual commercial plane could get his position accurately with respect to the field except by visual observation. In order to travel by plane one usually had first to make an intricate automobile trip from the center of the city out to a field in the country, and then, if there were fog or heavy rain, the trips would be cancelled. In spite of all these enormous handicaps a few courageous pioneers operated passenger lines and succeeded in giving acceptable service.

Of course the existing competition in the field of long-distance transportation had its disadvantages. Electrification of railroads had not proceeded far, and it was a common sight to see a great steam locomotive belching smoke and steam, radiating expensive heat units broadcast, and puffing away at its load. The driver leaned out of a side window in the rear, one end of him baked and the other end frozen. When the track curved the right way, he could see ahead. In order to start a train—for roller bearings were just appearing on rail-road cars—the locomotive would first back up to take up the slack in couplings, and then go ahead with a great bumping and crashing, much to the discomfort of passengers. Air con-ditioning was also in its infancy, and the atmosphere in the rear coaches of a train on a hot, dry summer day rocking across the country on uneven wood-ties and an unoiled roadbed may be imagined. The picture is more unattractive if one considers that sanitary arrangements were still somewhat barbarous.

Trains were enormously heavy and when well up to speed required a long distance in which to stop, for brakes were sim-ply pieces of cast iron pressed onto the steel wheels by air pres-sure. There were many highway crossings at grade, some of them actually without any automatic protection except a bell brought down from the old horse-drawn carriage days and inaudible to an automobile driver except when actually abreast, and not even then in a high wind. The locomotives carried shrieking whistles which blew almost constantly, much to the discomfiture of the countryside, but there were of course many grade crossing accidents.

Lunch for our professor was a ceremony of a sort. Nearly all the ingredients were of natural origin with very little in the way of synthetics or products of the biological industries. The foods were attractive in their way, but chosen almost at random, and served in circumstances that were somewhat appalling. To the discomforts noted in his office were added in the dining room a complete intermixture of the odors of all the several dishes and usually such incomplete ventilation that a blind man could have told the instant he entered the door of a restaurant. In fact he would not have had to depend upon his olfactory sense, for the dishes were of various glasses and ceramic materials (as well as in some cases the tables themselves), so that there was plenty of noise from the impact of hard materials.

On his return to the office, the same hectic round would continue. Perhaps in the afternoon the incessant clatter of typewriters would be especially annoying. For letters were written on typewriters, and there was a great deal of letter writing. That was the only way of practically conveying ideas outside of the archaic telephone and personal visit.

The library, to which our professor probably turned, was enormous. Long banks of shelves contained tons of books, and yet it was supposed to be a working library and not a museum. He had to paw over cards, thumb pages, and delve by the hour. It was time-wasting and exasperating indeed. Many of us well remember the amazing incredulity which greeted the first presentation of the unabridged dictionary on a square foot of film. The idea that one might have the contents of a thousand volumes located in a couple of cubic feet in a desk, so that by depressing a few keys one could have a given page instantly projected before him, was regarded as the wildest sort of fancy. This hesitation about accepting an idea, the basic soundness of which could have been tested by a little arithmetic, is worthy of more than passing notice. For the tenor of the age was to welcome new inventions and theories. In fact the man on the street was wont to visualize scientific triumphs as *faits accomplis* even as they were being hatched in the laboratory. He combined a simple credulity on some things, not erased even by the singeing of the Big Bull Market of the late 'Twenties, with a strange resistance to others. It seemed that the greater the technical difficulties which accomplished some really revolutionary proposition, the more casually the ordinary citizen accepted its consummation as being temporarily delayed but a fortnight or so.

Television was a case in point. To read the contemporary popular accounts one would suppose that the basic problem was solved at least once a month for several years. Yet the public seemed not to mind this crying of wolf, and quietly ignored simple analyses which showed that to transmit the image of a man's face in recognizable fashion would require 50 times the amount of communication channel adequate to transmit his voice. And when the progress of television proved to be exceedingly slow (like many other things which in the 'Thirties were asserted to be just around some corner), the layman was positive that the retardation was because of some corporation's machinations.

Somewhat the ordinary fellow of the 'Thirties, though he was by no means so witless as he deemed himself when he counted up his stock market losses in the earliest years of the decade, was quite muddled in his thinking process as seen from our present vantage point. He would, as I have said, readily accept the solution of such a complex thing as television to be imminent—as something he might find poking its way into his bedroom unawares on a bright Sunday morning. But he would consider a reasonable improvement in such an elementary thing as the arrangement of sleeping car space (it was really being tried by the railroads at the time) as incapable of realization for a couple of generations at least.

All about him he could see bridges, viaducts, steamships, engines, and so on, being built in hitherto unprecedented sizes. And, if some publicity agent issued an optimistic statement to the press that in the coming year they would be built twice as large again, he'd accept such a radical prediction with little emotion. Yet, when it was proposed to make it practicable for those who were neither too fat nor abnormally tall to undress in an upper berth, his reaction was likely to be expressed in the quaint vernacular of the day by some such expletive as "boloney" which, it seems, signified intense incredulity and an impatient skepticism.

As afternoon wore on the lights in the college buildings would be started. These were undoubtedly of the incandescent type in which a wire of tungsten in a gas-filled globe was heated by direct passage of current. About 95% of the energy furnished the lamp went into heating the room. The color was yellowish, and, as the sources were usually concentrated and small, there was formed a complex system of shadows. Lights would be forced on his attention again as he drove home, for there were no polarizers and the full beams of an approaching car would often strike in his eyes and temporarily blind him.

The professor's home had been built in position by the hand work of men of a dozen trades, who obtained materials by small-sized purchases from as many dealers, and cut and fit these materials on the site. It had meant an expensive capital outlay, and as much as a fifth of his income went into fixed charges on it. It had a cellar. This cellar, by way of explanation, was a large hole dug in the ground under the residence, and was a relic from the days when heating appliances required a cellar for their distribution pipes, which were so dirty that it was well to put them underground, and when thermal insulation was so imperfect that a ground floor of a house without a cellar was expected to be cold. Although these factors had ceased to be determining, the cellar with its expensive excavation had been continued as a sort of tradition.

Of course the house was immobile, and the idea of disassembling and transporting a house to a new site would have been considered quite radical. The heating plant was as crude as the one at the office, and even less efficient on account of its small size. Rooms were not soundproofed, hence many noises could be heard all over the house. This was particularly true of the plumbing fixtures, which made quite a racket. The floors were covered with fabric rugs, with their store of lint and dirt which was sometimes removed with a suction cleaner— incidentally the noisiest implement in the place. Some of the furniture was also fabric-covered and stuffed with natural horsehair as a sort of pseudo-elastic filling. It is not pleasant to contemplate in retrospect.

The excess of noise has been mentioned often in this review. One would think there would have been general protest, but the subject was only mildly mentioned in the contemporary newspaper. There was, in fact, some indication that the people generally liked to have a lot of noise about; at least they may have considered that its presence was conducive to intense activity in some manner. In no other way can we explain, for example, a turnstile which had been installed in large numbers in subways, and which was purposely arranged to give a loud clack whenever it was operated. This same peculiar love of noise was also exemplified in the contemporary music, which throws so much light on the strained mass psychology of the period.

The typical newspaper of the Nineteen-Thirties was a large affair, which reached truly ponderous proportions on Sunday. A single edition would then contain as much as 500 square feet of fine print. Of course no single individual could read all of this, and most people read much less than one per cent. Yet the forests were denuded—for paper was still made from pulp obtained from fair-sized trees—in order to print this mass of waste material. Bulky newspapers were one extravagance chargeable against the mania for advertising which had reached grotesque proportions. Nowhere were its extent and its methods more ludicrous than in the advertising of such things as cigarettes. A single phrase, usually quite absurd, would be repeated thousands of times in newspapers as well as on large placards placed by the roadsides (of course entirely out of harmony with the scenery), in trolley cars, and everywhere where space could be found. All this was done at enormous expense and without relief from the monotony of the repetition. One would expect there might have been a protest in the form of a boycott of any article thus intrusively offered, but, on the contrary, such methods apparently succeeded in their purpose. The cost of distribution to a public with so little true discrimination was, as we might expect, a heavy economic burden.

PECULIAR PASTIMES

Our professor, after his dinner, had many possible diversions. It is very likely that he participated in one of the peculiar popular pastimes which swept the country in sudden waves at about this period in history, exemplifying again the inexplicable mass psychology in force. One of these was a game called contract bridge, which persisted for several years and which was regarded with great seriousness by its devotees. It was participated in by both sexes, although it is hard to understand how their joint adherence to what was essentially a cult for mental exercise came about. Very complicated and artificial rules controlled the action, and the whole procedure must have been wearying, although it was indulged in by large sections of the population, sometimes to the substantial exclusion of all other mental activity. The force of mass opinion was so strong and individualism so repressed, that not to follow the public whim of the moment was to lose social caste. One of the strangest of these pastime manias occurred when over $100,000,000 was spent in the building of small gardens in which the participants knocked little white balls about among various obstacles. This lasted, as would be expected, for only a few months.

An epidemic of jig-saw puzzles also swept the land. Though they were harmless in themselves, the distress of elderly people and children alike when some piece or pieces dropped to the floor and became hidden in some out-of-the-way place, was apt to be expressed in fitful bursts of temper.

It is little wonder that under such hectic conditions many professors developed otherwise unaccountable tendencies, and that the science of the times was disjointed, heterogeneous, and very much an opportunist affair. There should be the greatest of admiration for one who could do any constructive thinking at all when thus badgered about during the Nineteen-Thirties.

Today, quite correctly, we realize that all of the desirable modifications of our natural environment, which are possible through simple mechanical means, have been accomplished. Even 20 years ago there were significant indications that the door to progress along such lines was fast closing. About that time the standing Congressional Committee, appointed to consider the scope of the world's scientific and technical progress with special reference to its bearing upon matters to be discussed at a proposed conference to be convened for the further discussion of inter-governmental debts, was able to include in its preliminary report the convincing statement that everything worth while had been done.

What a contrast to the Nineteen-Thirties! Then it should have been possible, even by the most rudimentary of analyses, to disclose attractive avenues for constructive effort.

It may be asked why, with all this opportunity, we had to wait so long for the obvious. It could not have been because of a lack of leisure; for there were emphatic contemporary complaints that leisure engendered by the quickening of production had become both burdensome and notoriously unwelcome. Perhaps it was ascribable in a measure to the prevailing social code which then forced all men to dress alike and, to some extent, to think alike. Or, it may have been that the pressure of advertising propaganda had induced a mass psychology which led people to believe they had arrived at some sort of mechanical Utopia with which they were in duty bound to be content.

These few suggestions are typical of the myriad of hypotheses contained in the extensive literature dealing with this quite amazing decade in the history of the Republic. From all the thought given to the subject but one solitary fact emerges: the Nineteen-Thirties still remain inscrutable.

2: AS WE MAY THINK

World War II was not a scientist's war; it was a war in which all had a part. The scientists, burying their old professional competition in the demand of a common cause, shared greatly and learned much. It was exhilarating to work in effective partnership. Now, for many, this appears to be approaching an end. What are the scientists to do next?

For the biologists, and particularly for the medical scientists, there can be little indecision, for their war work hardly required them to leave the old paths. Many indeed were able to carry on their war research in their familiar peacetime laboratories. Their objectives remain much the same.

It is the physicists who were thrown most violently off stride, who left academic pursuits for the making of strange destructive gadgets, who had to devise new methods for their unanticipated assignments. They did their part on the devices that made it possible to turn back the enemy. They worked in combined effort with the physicists of our allies. They felt within themselves the stir of achievement. They were part of a great team. Now one asks where they will find objectives worthy of their best.

THE BENEFITS OF SCIENCE

Of what lasting benefit has been man's use of science and of the new instruments which his research brought into existence? First, they have increased his control of his material environment. They have improved his food, his clothing, his shelter; they have increased his security and released him partly from the bondage of bare existence. They have given him increased knowledge of his own biological processes so that he has had

16

a progressive freedom from disease and an increased span of life. They are illuminating the interactions of his physiological and psychological functions, giving the promise of an improved mental health.

Science has provided the swiftest communication between individuals; it has provided a record of ideas and has enabled man to manipulate and to make extracts from that record so that knowledge evolves and endures throughout the life of a race rather than that of an individual.

There is a growing mountain of research. But there is increased evidence that we are being bogged down today as specialization extends. The investigator is staggered by the findings and conclusions of thousands of other workers—many of which he cannot find time to grasp, much less to remember, as they appear. Yet specialization becomes increasingly necessary for progress, and the effort to bridge between disciplines is correspondingly superficial.

Professionally our methods of transmitting and reviewing the results of research are generations old and by now are totally inadequate for their purpose. If the aggregate time spent in writing scholarly works and in reading them could be evaluated, the ratio between these amounts of time might well be startling. Those who conscientiously attempt to keep abreast of current thought, even in restricted fields, by close and continuous reading might well shy away from an examination calculated to show how much of the previous month's efforts could be produced on call. Mendel's concept of the laws of genetics was lost to the world for a generation because his publication did not reach the few who were capable of grasping and extending it; and this sort of catastrophe is undoubtedly being repeated all about us, as truly significant attainments become lost in the mass of the inconsequential.

The difficulty seems to be, not so much that we publish unduly in view of the extent and variety of present-day interests, but rather that publication has been extended far beyond our present ability to make real use of the record. The summation of human experience is being expanded at a prodigious rate, and the means we use for threading through the consequent maze to the momentarily important item is the same as was used in the days of square-rigged ships.

But there are signs of a change as new and powerful instrumentalities come into use. Photocells capable of seeing things in a physical sense, advanced photography which can record what is seen or even what is not, thermionic tubes capable of controlling potent forces under the guidance of less power than a mosquito uses to vibrate his wings, cathode ray tubes rendering visible an occurrence so brief that by comparison a microsecond is a long time, relay combinations which will carry out involved sequences of movements more reliably than any human operator and thousands of times as fast—there are plenty of mechanical aids with which to effect a transformation in scientific records.

Two centuries ago Leibnitz invented a calculating machine which embodied most of the essential features of recent keyboard devices, but it could not then come into use. The economics of the situation were against it: the labor involved in constructing it, before the days of mass production, exceeded the labor to be saved by its use, since all it could accomplish could be duplicated by sufficient use of pencil and paper. Moreover, it would have been subject to frequent breakdown, so that it could not have been depended upon; for at that time and long after, complexity and unreliability were synonymous.

Babbage, even with remarkably generous support for his time, could not produce his great arithmetical machine. His idea was sound enough, but construction and maintenance costs were then too heavy. Had a Pharaoh been given detailed and explicit designs of an automobile, and had he understood them completely, it would have taxed the resources of his kingdom to fashion the thousands of parts for a single car, and that car would have broken down on the first trip to Giza.

Machines with interchangeable parts can now be constructed with great economy of effort. In spite of much complexity, they perform reliably. Witness the humble typewriter, or the movie camera, or the automobile. Electrical contacts have ceased to stick when thoroughly understood. Note the automatic telephone exchange, which has hundreds of thousands of such contacts, and yet is reliable. A spider web of metal, sealed in a thin glass container, a wire heated to brilliant glow, in short, the thermionic tube of radio sets, is made by the hundred million, tossed about in packages, plugged into sockets—and it works! Its gossamer parts, the precise location and alignment involved in its construction, would have occupied a master craftsman of the guild for months; now it is built for thirty cents. The world has arrived at an age of cheap complex devices of great reliability; and something is bound to come of it.

RECORDING DEVICES

A record, if it is to be useful to science, must be continuously extended, it must be stored, and above all it must be consulted. Today we make the record conventionally by writing and photography, followed by printing; but we also record on film, on wax disks, and on magnetic wires. Even if utterly new recording procedures do not appear, these present ones are certainly in the process of modification and extension.

Certainly progress in photography is not going to stop. Faster material and lenses, more automatic cameras, finer-grained sensitive compounds to allow an extension of the mini-camera idea, are all imminent. Let us project this trend ahead to a logical, if not inevitable, outcome. The camera hound of the future wears on his forehead a lump a little larger than a walnut. It takes pictures 3 millimeters square, later to be projected or enlarged, which after all involves only a factor of 10 beyond present practice. The lens is of universal focus, down to any distance accommodated by the unaided eye, simply because it is of short focal length. There is a built-in photocell on the walnut such as we now have on at least one camera, which automatically adjusts exposure for a wide range of illumination. There is film in the walnut for a hundred exposures, and the spring for operating its shutter and shifting its film is wound once for all when the film clip is inserted. It produces its result in full color. It may well be stereoscopic, and record with two spaced glass eyes, for striking improvements in stereoscopic technique are just around the corner.

The cord which trips its shutter may reach down a man's sleeve within easy reach of his fingers. A quick squeeze, and the picture is taken. On a pair of ordinary glasses is a square of fine lines near the top of one lens, where it is out of the way of ordinary vision. When an object appears in that square, it is lined up for its picture. As the scientist of the future moves about the laboratory or the field, every time he looks at something worthy of the record, he trips the shutter and in it goes, without even an audible click. Is this all fantastic? The only fantastic thing about it is the idea of making as many pictures as would result from its use.

Will there be dry photography? It is already here in two forms. When Brady made his Civil War pictures, the plate had to be wet at the time of exposure. Now it has to be wet during development instead. In the future perhaps it need not be wetted at all. There have long been films impregnated with diazo dyes which form a picture without development, so that it is already there as soon as the camera has been operated. An exposure to ammonia gas destroys the unexposed dye, and the picture can then be taken out into the light and examined. The process is now slow, but someone may speed it up, and it has no grain difficulties such as now keep photographic researchers busy. Often it would be advantageous to be able to snap the camera and to look at the picture immediately.

Another process now in use is also slow, and more or less clumsy. For fifty years impregnated papers have been used which turn dark at every point where an electrical contact touches them, by reason of the chemical change thus produced in an iodine compound included in the paper. They have been used to make records, for a pointer moving across them can leave a trail behind. If the electrical potential on the pointer is varied as it moves, the line becomes light or dark in accordance with the potential.

This scheme is now used in facsimile transmission. The pointer draws a set of closely spaced lines across the paper one after another. As it moves, its potential is varied in accordance with a varying current received over wires from a distant station, where these variations are produced by a photocell which is similarly scanning a picture. At every instant the darkness of the line being drawn is made equal to the darkness of the point on the picture being observed by the photocell. Thus, when the whole picture has been covered, a replica appears at the receiving end.

A scene itself can be just as well looked over line by line by the photocell in this way as can a photograph of the scene. This whole apparatus constitutes a camera, with the added feature, which can be dispensed with if desired, of making its picture at a distance. It is slow, and the picture is poor in detail. Still, it does give another process of dry photography, in which the picture is finished as soon as it is taken.

It would be a brave man who would predict that such a process will always remain clumsy, slow, and faulty in detail. Television equipment today transmits sixteen reasonably good pictures a second, and it involves only two essential differences from the process described above. For one, the record is made by a moving beam of electrons rather than a moving pointer, for the reason that an electron beam can sweep across the picture very rapidly indeed. The other difference involves merely the use of a screen which glows momentarily when the electrons hit, rather than a chemically treated paper or film which is permanently altered. This speed is necessary in television, for motion pictures rather than stills are the object.

Use chemically treated film in place of the glowing screen, allow the apparatus to transmit one picture only rather than a succession, and a rapid camera for dry photography results. The treated film needs to be far faster in action than present examples, but it probably could be. More serious is the objection that this scheme would involve putting the film inside a vacuum chamber, for electron beams behave normally only in such a rarefied environment. This difficulty could be avoided by allowing the electron beam to play on one side of a partition, and by pressing the film against the other side, if this partition were such as to allow the electrons to go through perpendicular to its surface, and to prevent them from spreading out sideways. Such partitions, in crude form, could certainly be constructed, and they will hardly hold up the general development.

Like dry photography, microphotography still has a long way to go. The basic scheme of reducing the size of the record, and examining it by projection rather than directly, has possibilities too great to be ignored. The combination of optical projection and photographic reduction is already producing some results in microfilm for scholarly purposes, and the potentialities are highly suggestive. Today, with microfilm, reductions by a linear factor of 20 can be employed and still produce full clarity when the material is re-enlarged for examination. The limits are set by the graininess of the film, the excellence of the optical system, and the efficiency of the light sources employed. All of these are rapidly improving.

Assume a linear ratio of 100 for future use. Consider film of the same thickness as paper, although thinner film will certainly be usable. Even under these conditions there would be a total factor of 10,000 between the bulk of the ordinary record on books, and its microfilm replica. The *Encyclopædia Britannica* could be reduced to the volume of a matchbox. A library of a million volumes could be compressed into one end of a desk. If the human race has produced since the invention of movable type a total record, in the form of magazines, newspapers, books, tracts, advertising blurbs, correspondence, having a volume corresponding to a billion books, the whole affair, assembled and compressed, could be lugged off in a moving van. Mere compression, of course, is not enough; one needs not only to make and store a record but also to be able to consult it, and this aspect of the matter comes later. Even the modern great library is not generally consulted; it is nibbled at by a few.

Compression is important, however, when it comes to costs. The material for the microfilm *Britannica* would cost a nickel, and it could be mailed anywhere for a cent. What would it cost to print a million copies? To print a sheet of newspaper, in a large edition, costs a small fraction of a cent. The entire material of the *Britannica* in reduced microfilm form would go on a sheet eight and one-half by eleven inches. Once it is available, with the photographic reproduction methods of the future, duplicates in large quantities could probably be turned out for a cent apiece beyond the cost of materials. The preparation of the original copy? That introduces the next aspect of the subject.

RECORDING PROCESSES

To make the record, we now push a pencil or tap a typewriter. Then comes the process of digestion and correction, followed by an intricate process of typesetting, printing, and distribution. To consider the first stage of the procedure, will the author of the future cease writing by hand or typewriter and talk directly to the record? He does so indirectly, by talking to a stenographer or a wax cylinder; but the elements are all present if he wishes to have his talk directly produce a typed record. All he needs to do is to take advantage of existing mechanisms and to alter his language.

At a recent World Fair a machine called a Voder was shown. A girl stroked its keys and it emitted recognizable speech. No human vocal chords entered into the procedure at any point; the keys simply combined some electrically produced vibrations and passed these on to a loud-speaker. In the Bell Laboratories there is the converse of this machine, called a Vocoder. The loudspeaker is replaced by a microphone, which picks up sound. Speak to it, and the corresponding keys move. This may be one element of the postulated system.

The other element is found in the stenotype, that somewhat disconcerting device encountered usually at public meetings. A girl strokes its keys languidly and looks about the room and sometimes at the speaker with a disquieting gaze. From it emerges a typed strip which records in a phonetically simplified language a record of what the speaker is supposed to have said. Later this strip is retyped into ordinary language, for in its nascent form it is intelligible only to the initiated. Combine these two elements, let the Vocoder run the stenotype, and the result is a machine which types when talked to.

Our present languages are not especially adapted to this sort of mechanization, it is true. It is strange that the inventors of universal languages have not seized upon the idea of producing one which better fitted the technique for transmitting and recording speech. Mechanization may yet force the issue, especially in the scientific field; whereupon scientific jargon would become still less intelligible to the layman.

One can now picture a future investigator in his laboratory. His hands are free, and he is not anchored. As he moves about and observes, he photographs and comments. Time is automatically recorded to tie the two records together. If he goes into the field, he may be connected by radio to his recorder. As he ponders over his notes in the evening, he again talks his comments into the record. His typed record, as well as his photographs, may both be in miniature, so that he projects them for examination.

Much needs to occur, however, between the collection of data and observations, the extraction of parallel material from the existing record, and the final insertion of new material into the general body of the common record. For mature thought there is no mechanical substitute. But creative thought and essentially repetitive thought are very different things. For the latter there are, and may be, powerful mechanical aids.

Adding a column of figures is a repetitive thought process, and it was long ago properly relegated to the machine. True, the machine is sometimes controlled by a keyboard, and thought of a sort enters in reading the figures and poking the corresponding keys, but even this is avoidable. Machines which will read typed figures by photocells and then depress the corresponding keys may come from combinations of photocells for scanning the type, electric circuits for sorting the consequent variations, and relay circuits for interpreting the result into the action of solenoids to pull the keys down.

All this complication is needed because of the clumsy way in which we have learned to write figures. If we recorded them positionally, simply by the configuration of a set of dots on a card, the automatic reading mechanism would become comparatively simple. In fact, if the dots are holes, we have the punched-card machine long ago produced by Hollerith for the purposes of the census, and now used throughout business. Some types of complex businesses could hardly operate without these machines.

Adding is only one operation. To perform arithmetical computation involves also subtraction, multiplication, and division, and in addition some method for temporary storage of results, removal from storage for further manipulation, and recording of final results by printing. Machines for these purposes are now of two types: keyboard machines for accounting and the like, manually controlled for the insertion of data, and usually automatically controlled as far as the sequence of operations is concerned; and punched-card machines in which separate operations are usually delegated to a series of machines, and the cards then transferred bodily from one to another. Both forms are very useful; but as far as complex computations are concerned, both are still in embryo.

Rapid electrical counting appeared soon after the physicists found it desirable to count cosmic rays. For their own purposes the physicists promptly constructed thermionic-tube equipment capable of counting electrical impulses at the rate of 100,000 a second. The advanced arithmetical machines of the future will be electrical in nature, and they will perform at 100 times present speeds, or more.

Moreover, they will be far more versatile than present commercial machines, so that they may readily be adapted for a wide variety of operations. They will be controlled by a control card or film, they will select their own data and manipulate them in accordance with instructions thus inserted, they will perform complex arithmetical computations at exceedingly high speeds, and they will record results in such form as to be readily available for distribution or for later further manipulation. Such machines will have enormous appetites. One of them will take instructions and data from a whole roomful of girls armed with simple keyboard punches, and will deliver sheets of computed results every few minutes. There will always be plenty of things to compute in the detailed affairs of millions of people doing complicated things.

SPECIAL MACHINES

The repetitive processes of thought are not confined, however, to matters of arithmetic and statistics. In fact, every time one combines and records facts in accordance with established logical processes, the creative aspect of thinking is concerned only with the selection of the data and the process to be employed, and the manipulation thereafter is repetitive in nature and hence a fit matter to be relegated to the machines. Not so much has been done along these lines, beyond the bounds of arithmetic, as might be done, primarily because of the economics of the situation. The needs of business, and the extensive market obviously waiting, assured the advent of mass-produced arithmetical machines just as soon as production methods were sufficiently advanced.

With machines for advanced analysis no such situation existed; for there was and is no extensive market; the users of advanced methods of manipulating data are a very small part of the population. There are, however, machines for solving differential equations—and functional and integral equations, for that matter. There are many special machines, such as the harmonic synthesizer which predicts the tides. There will be many more, appearing certainly first in the hands of the scientist and in small numbers.

If scientific reasoning were limited to the logical processes of arithmetic, we should not get far in our understanding of the physical world. One might as well attempt to grasp the game of poker entirely by the use of the mathematics of probability. The abacus, with its beads strung on parallel wires, led the Arabs to positional numeration and the concept of zero many centuries before the rest of the world; and it was a useful tool—so useful that it still exists.

It is a far cry from the abacus to the modern keyboard accounting machine. It will be an equal step to the arithmetical machine of the future. But even this new machine will not take the scientist where he needs to go. Relief must be secured from laborious detailed manipulation of higher mathematics as well, if the users of it are to free their brains for something more than repetitive detailed transformations in accordance with established rules. A mathematician is not a man who can readily manipulate figures; often he cannot. He is not even a man who can readily perform the transformations of equations by the use of calculus. He is primarily an individual who is skilled in the use of symbolic logic on a high plane, and especially he is a man of intuitive judgment in the choice of the manipulative processes he employs.

All else he should be able to turn over to his mechanism, just as confidently as he turns over the propelling of his car to the intricate mechanism under the hood. Only then will mathematics be practically effective in bringing the growing knowledge of atomistics to the useful solution of the advanced problems of chemistry, metallurgy, and biology. For this reason there will come more machines to handle advanced mathematics for the scientist. Some of them will be sufficiently bizarre to suit the most fastidious connoisseur of the present artifacts of civilization.

SELECTION TECHNIQUES

The scientist, however, is not the only person who manipulates data and examines the world about him by the use of logical processes, although he sometimes preserves this appearance by adopting into the fold anyone who becomes logical, much in the manner in which a British labor leader is elevated to knighthood. Whenever logical processes of thought are employed—that is, whenever thought for a time runs along an accepted groove—there is an opportunity for the machine. Formal logic used to be a keen instrument in the hands of the teacher in his trying of students' souls. It is readily possible to construct a machine which will manipulate premises in accordance with formal logic, simply by the clever use of relay circuits. Put a set of premises into such a device and turn the crank, and it will readily pass out conclusion after conclusion, all in accordance with logical law, and with no more slips than would be expected of a keyboard adding machine.

Logic can become enormously difficult, and it would undoubtedly be well to produce more assurance in its use. The machines for higher analysis have usually been equation solvers. Ideas are beginning to appear for equation transformers, which will rearrange the relationship expressed by an equation in accordance with strict and rather advanced logic. Progress is inhibited by the exceedingly crude way in which mathematicians express their relationships. They employ a symbolism which grew like Topsy and has little consistency; a strange fact in that most logical field.

A new symbolism, probably positional, must apparently precede the reduction of mathematical transformations to machine processes. Then, on beyond the strict logic of the mathematician, lies the application of logic in everyday affairs. We may some day click off arguments on a machine with the same assurance that we now enter sales on a cash register. But the machine of logic will not look like a cash register, even the streamlined model.

So much for the manipulation of ideas and their insertion into the record. Thus far we seem to be worse off than before —for we can enormously extend the record; yet even in its present bulk we can hardly consult it. This is a much larger matter than merely the extraction of data for the purposes of scientific research; it involves the entire process by which man profits by his inheritance of acquired knowledge. The prime action of use is selection, and here we are halting indeed. There may be millions of fine thoughts, and the account of the experience on which they are based, all encased within stone walls of acceptable architectural form; but if the scholar can get at only one a week by diligent search, his syntheses are not likely to keep up with the current scene.

Selection, in this broad sense, is a stone adze in the hands of a cabinetmaker. Yet, in a narrow sense and in other areas, something has already been done mechanically on selection. The personnel officer of a factory drops a stack of a few thousand employee cards into a selecting machine, sets a code in accordance with an established convention, and produces in a short time a list of all employees who live in Trenton and know Spanish. Even such devices are much too slow when it comes, for example, to matching a set of fingerprints with one of five million on file. Selection devices of this sort will soon be speeded up from their present rate of reviewing data at a few hundred a minute. By the use of photocells and microfilm they will survey items at the rate of a thousand a second, and will print out duplicates of those selected.

This process, however, is simple selection: it proceeds by examining in turn every one of a large set of items, and by picking out those which have certain specified characteristics. There is another form of selection best illustrated by the automatic telephone exchange. You dial a number and the machine selects and connects just one of a million possible stations. It does not run over them all. It pays attention only to a class given by a first digit, then only to a subclass of this given by the second digit, and so on; and thus proceeds rapidly and almost unerringly to the selected station. It requires a few seconds to make the selection, although the process could be speeded up if increased speed were economically warranted. If necessary, it could be made extremely fast by substituting thermionic-tube switching for mechanical switching, so that the full selection could be made in one one-hundredth of a second. No one would wish to spend the money necessary to make this change in the telephone system, but the general idea is applicable elsewhere.

Take the prosaic problem of the great department store. Every time a charge sale is made, there are a number of things to be done. The inventory needs to be revised, the salesman needs to be given credit for the sale, the general accounts need an entry, and, most important, the customer needs to be charged. A central records device has been developed in which much of this work is done conveniently. The salesman places on a stand the customer's identification card, his own card, and the card taken from the article sold—all punched cards. When he pulls a lever, contacts are made through the holes, machinery at a central point makes the necessary computations and entries, and the proper receipt is printed for the salesman to pass to the customer.

But there may be ten thousand charge customers doing business with the store, and before the full operation can be completed someone has to select the right card and insert it at the central office. Now rapid selection can slide just the proper card into position in an instant or two, and return it afterward. Another difficulty occurs, however. Someone must read a total on the card, so that the machine can add its computed item to it. Conceivably the cards might be of the dry photography type I have described. Existing totals could then be read by photocell, and the new total entered by an electron beam.

The cards may be in miniature, so that they occupy little space. They must move quickly. They need not be transferred far, but merely into position so that the photocell and recorder can operate on them. Positional dots can enter the data. At the end of the month a machine can readily be made to read these and to print an ordinary bill. With tube selection, in which no mechanical parts are involved in the switches, little time need be occupied in bringing the correct card into use—a second should suffice for the entire operation. The whole record on the card may be made by magnetic dots on a steel sheet if desired, instead of dots to be observed optically, following the scheme by which Poulsen long ago put speech on a magnetic wire. This method has the advantage of simplicity and ease of erasure. By using photography, however, one can arrange to project the record in enlarged form, and at a distance by using the process common in television equipment.

One can consider rapid selection of this form and distant projection for other purposes. To be able to key one sheet of a million before an operator in a second or two, with the possibility of then adding notes thereto, is suggestive in many ways. It might even be of use in libraries, but that is another story. At any rate, there are now some interesting combinations possible. One might, for example, speak to a microphone, in the manner described in connection with the speech-controlled typewriter, and thus make his selections. It would certainly beat the usual file clerk.

MEMEX INSTEAD OF INDEX

The real heart of the matter of selection, however, goes deeper than a lag in the adoption of mechanisms by libraries, or a lack of development of devices for their use. Our ineptitude in getting at the record is largely caused by the artificiality of systems of indexing. When data of any sort are placed in storage, they are filed alphabetically or numerically, and information is found (when it is) by tracing it down from subclass to subclass. It can be in only one place, unless duplicates are used; one has to have rules as to which path will locate it, and the rules are cumbersome. Having found one item, moreover, one has to emerge from the system and re-enter on a new path.

The human mind does not work that way. It operates by

association. With one item in its grasp, it snaps instantly to the next that is suggested by the association of thoughts, in accordance with some intricate web of trails carried by the cells of the brain. It has other characteristics, of course; trails that are not frequently followed are prone to fade, items are not fully permanent, memory is transitory. Yet the speed of action, the intricacy of trails, the detail of mental pictures, is awe-inspiring beyond all else in nature.

Man cannot hope fully to duplicate this mental process artificially, but he certainly ought to be able to learn from it. In minor ways he may even improve, for his records have relative permanency. The first idea, however, to be drawn from the analogy concerns selection. Selection by association, rather than by indexing, may yet be mechanized. One cannot hope thus to equal the speed and flexibility with which the mind follows an associative trail, but it should be possible to beat the mind decisively in regard to the permanence and clarity of the items resurrected from storage.

Consider a future device for individual use, which is a sort of mechanized private file and library. It needs a name, and, to coin one at random, "memex" will do. A memex is a device in which an individual stores all his books, records, and communications, and which is mechanized so that it may be consulted with exceeding speed and flexibility. It is an enlarged intimate supplement to his memory.

It consists of a desk, and while it can presumably be operated from a distance, it is primarily the piece of furniture at which he works. On the top are slanting translucent screens, on which material can be projected for convenient reading. There is a keyboard, and sets of buttons and levers. Otherwise it looks like an ordinary desk.

In one end is the stored material. The matter of bulk is well taken care of by improved microfilm. Only a small part of the interior of the memex is devoted to storage, the rest to mechanism. Yet if the user inserted 5,000 pages of material a day it would take him hundreds of years to fill the repository, so he can be profligate and enter material freely.

Most of the memex contents are purchased on microfilm ready for insertion. Books of all sorts, pictures, current periodicals, newspapers, are thus obtained and dropped into place. Business correspondence takes the same path. And there is provision for direct entry. On the top of the memex is a transparent platen. On this are placed longhand notes, photographs, memoranda, all sorts of things. When one is in place, the depression of a lever causes it to be photographed onto the next blank space in a section of the memex film, dry photography being employed.

There is, of course, provision for consultation of the record by the usual scheme of indexing. If the user wishes to consult a certain book, he taps its code on the keyboard, and the title page of the book promptly appears before him, projected onto one of his viewing positions. Frequently-used codes are mnemonic, so that he seldom consults his code book; but when he does, a single tap of a key projects it for his use. Moreover, he has supplemental levers. On deflecting one of these levers to the right he runs through the book before him, each page in turn being projected at a speed which just allows a recognizing glance at each. If he deflects it further to the right, he steps through the book 10 pages at a time; still further at 100 pages at a time. Deflection to the left gives him the same control backwards.

A special button transfers him immediately to the first page of the index. Any given book of his library can thus be called up and consulted with far greater facility than if it were taken from a shelf. As he has several projection positions, he can leave one item in position while he calls up another. He can add marginal notes and comments, taking advantage of one possible type of dry photography, and it could even be arranged so that he can do this by a stylus scheme, such as is now employed in the telautograph seen in railroad waiting rooms, just as though he had the physical page before him.

ENDLESS TRAILS

All this is conventional, except for the projection forward of present-day mechanisms and gadgetry. It affords an immediate step, however, to associative indexing, the basic idea of which is a provision whereby any item may be caused at will to select immediately and automatically another. This is the essential feature of the memex. The process of tying two items together is the important thing.

When the user is building a trail, he names it, inserts the name in his code book, and taps it out on his keyboard. Before him are the two items to be joined, projected onto adjacent viewing positions. At the bottom of each there are a number of blank code spaces, and a pointer is set to indicate one of these on each item. The user taps a single key, and the items are permanently joined. In each code space appears the code word. Out of view, but also in the code space, is inserted a set of dots for photocell viewing; and on each item these dots by their positions designate the index number of the other item.

Thereafter, at any time, when one of these items is in view, the other can be instantly recalled merely by tapping a button below the corresponding code space. Moreover, when numerous items have been thus joined together to form a trail, they can be reviewed in turn, rapidly or slowly, by deflecting a lever like that used for turning the pages of a book. It is exactly as though the physical items had been gathered together from widely separated sources and bound together to form a new book. It is more than this, for any item can be joined into numerous trails.

The owner of the memex, let us say, is interested in the origin and properties of the bow and arrow. Specifically he i studying why the short Turkish bow was apparently superior to the English long bow in the skirmishes of the Crusades. He has dozens of possibly pertinent books and articles in his memex. First he runs through an encyclopedia, finds an interesting but sketchy article, leaves it projected. Next, in a history, he finds another pertinent item, and ties the two together. Thus he goes, building a trail of many items. Occasionally he inserts a comment of his own, either linking it into the main trail or joining it by a side trail to a particular item. When it becomes evident that the elastic properties of available materials had a great deal to do with the bow, he branches off on a side trail which takes him through textbooks on elasticity and tables of physical constants. He inserts a page of longhand analysis of his own. Thus he builds a trail of his interest through the maze of materials available to him.

And his trails do not fade. Several years later, his talk with a friend turns to the queer ways in which a people resist innovations, even of vital interest. He has an example, in the fact that the outranged Europeans still failed to adopt the Turkish bow. In fact he has a trail on it. A touch brings up the code book. Tapping a few keys projects the head of the trail. A lever runs through it at will, stopping at interesting items, going off on side excursions. It is an interesting trail, pertinent to the discussion. So he sets a reproducer in action, photographs the whole trail out, and passes it to his friend for insertion in his own memex, there to be linked into the more general trail.

HORIZONS UNLIMITED

Wholly new forms of encyclopedias will appear, ready-made with a mesh of associative trails running through them, ready to be dropped into the memex and there amplified. The lawyer has at his touch the associated opinions and decisions of his whole experience, and of the experience of friends and authorities. The patent attorney has on call the millions of issued patents, with familiar trails to every point of his client's interest. The physician, puzzled by a patient's reactions, strikes the trail established in studying an earlier similar case, and runs rapidly through analogous case histories, with side references to the classics for the pertinent anatomy and histology. The chemist, struggling with the synthesis of an organic compound, has all the chemical literature before him in his laboratory, with trails following the analogies of compounds, and side trails to their physical and chemical behavior.

The historian, with a vast chronological account of a people, parallels it with a skip trail which stops only on the salient items, and can follow at any time contemporary trails which lead him all over civilization at a particular epoch. There is a new profession of trail blazers, those who find delight in the task of establishing useful trails through the enormous mass of the common record. The inheritance from the master becomes, not only his additions to the world's record, but for his disciples the entire scaffolding by which they were erected.

Thus science may implement the ways in which man produces, stores, and consults the record of the race. It might be striking to outline the instrumentalities of the future more spectacularly, rather than to stick closely to methods and elements now known and undergoing rapid development, as has been done here. Technical difficulties of all sorts have been ignored, certainly, but also ignored are means as yet unknown which may come any day to accelerate technical progress as violently as did the advent of the thermionic tube. In order that the picture may not be too commonplace, by reason of sticking to present-day patterns, it may be well to mention one such possibility, not to prophesy but merely to suggest, for prophecy based on extension of the known has substance, while prophecy founded on the unknown is only a doubly involved guess.

All our steps in creating or absorbing material of the record proceed through one of the senses—the tactile when we touch keys, the oral when we speak or listen, the visual when we read. Is it not possible that some day the path may be established more directly?

We know that when the eye sees, all the consequent information is transmitted to the brain by means of electrical vibrations in the channel of the optic nerve. This is an exact analogy with the electrical vibrations which occur in the cable of a television set: they convey the picture from the photocells which see it to the radio transmitter from which it is broadcast. We know further that if we can approach that cable with the proper instruments, we do not need to touch it; we can pick up those vibrations by electrical induction and thus discover and reproduce the scene which is being transmitted, just as a telephone wire may be tapped for its message.

The impulses which flow in the arm nerves of a typist convey to her fingers the translated information which reaches her eye or ear, in order that the fingers may be caused to strike the proper keys. Might not these currents be intercepted, either in the original form in which information is conveyed to the brain, or in the marvelously metamorphosed form in which they then proceed to the hand?

By bone conduction we already introduce sounds into the nerve channels of the deaf in order that they may hear. Is it not possible that we may learn to introduce them without the present cumbersomeness of first transforming electrical vibrations to mechanical ones, which the human mechanism promptly transforms back to the electrical form? With a couple of electrodes on the skull the encephalograph now produces pen-and-ink traces which bear some relation to the electrical phenomena going on in the brain itself. True, the record is unintelligible, except as it points out certain gross misfunctioning of the cerebral mechanism; but who would now place bounds on where such a thing may lead?

In the outside world, all forms of intelligence, whether of sound or sight, have been reduced to the form of varying currents in an electric circuit in order that they may be transmitted. Inside the human frame exactly the same sort of process occurs. Must we always transform to mechanical movements in order to proceed from óne electrical phenomenon to another? It is a suggestive thought, but it hardly warrants prediction without losing touch with reality and immediateness.

Presumably man's spirit should be elevated if he can better review his shady past and analyze more completely and objectively his present problems. He has built a civilization so complex that he needs to mechanize his records more fully if he is to push his experiment to its logical conclusion and not merely become bogged down part way there by overtaxing his limited memory. His excursions may be more enjoyable if he can reacquire the privilege of forgetting the manifold things he does not need to have immediately at hand, with some assurance that he can find them again if they prove important.

The applications of science have built man a well-supplied house, and are teaching him to live healthily therein. They have enabled him to throw masses of people against one another with cruel weapons. They may yet allow him truly to encompass the great record and to grow in the wisdom of race experience. He may perish in conflict before he learns to wield that record for his true good. Yet, in the application of science to the needs and desires of man, this would seem to be a singularly unfortunate stage at which to terminate the process, or to lose hope as to the outcome.

3: A PROGRAM FOR TOMORROW

In a letter dated November 17, 1944, President Roosevelt requested my recommendations on the following points:

(1) What can be done, consistent with military security, and with the prior approval of the military authorities, to make known to the world as soon as possible the contributions which have been made during our war effort to scientific knowledge?

(2) With particular reference to the war of science against disease, what can be done now to organize a program for continuing in the future the work which has been done in medicine and related sciences?

(3) What can the Government do now and in the future to aid research activities by public and private organizations?

(b) Can an effective program be proposed for discovering and developing scientific talent in American youth so that the continuing future of scientific research in this country may be assured on a level comparable to what has been done during the war?

It is clear from President Roosevelt's letter that in speaking of science he had in mind the natural sciences, including biology and medicine, and I have so interpreted his questions. Progress in other fields, such as the social sciences and the humanities, is likewise important; but the program for science presented in my report warrants immediate attention.

In seeking answers to President Roosevelt's questions I have had the assistance of distinguished committees specially qualified to advise in respect to these subjects. The committees have given these matters the serious attention they deserve; indeed, they have regarded this as an opportunity to participate in shaping the policy of the country with reference to scientific research. They have had many meetings and have submitted formal reports. I have been in close touch with the work of the committees and with their members throughout. I have examined all of the data they assembled and the suggestions they submitted on the points raised in President Roosevelt's letter.

Although the report which I drew up is my own, the facts, conclusions, and recommendations are based on the findings of the committees which have studied these questions.

A single mechanism for implementing the recommendations of the several committees is essential. In proposing such a mechanism I have departed somewhat from the specific recommendations of the committees, but I have since been assured that the plan I am proposing is fully acceptable to the committee members.

The pioneer spirit is still vigorous within this nation. Science offers a largely unexplored hinterland for the pioneer who has the tools for his task. The rewards of such exploration for both the nation and the individual are great. Scientific progress is one essential key to our security as a nation, to our better health, to more jobs, to a higher standard of living, and to our cultural progress.

NATIONAL PROGRESS

We all know how much the new drug, penicillin, has meant to our grievously wounded men on the grim battlefronts—the countless lives it has saved—the incalculable suffering which its use has prevented. Science and the great practical genius of this nation made this achievement possible.

Some of us know the vital role which radar has played in bringing the United Nations to victory over Nazi Germany and in driving the Japanese steadily back from their island bastions. Again it was painstaking scientific research over many years that made radar possible.

What we often forget are the millions of pay envelopes on a peacetime Saturday night which are filled because new products and new industries have provided jobs for countless Americans. Science made that possible, too.

In 1939 millions of people were employed in industries which did not even exist at the close of the last war—radio, air conditioning, rayon and other synthetic fibers, and plastics are examples of the products of these industries. But these things do not mark the end of progress—they are but the beginning if we make full use of our scientific resources. New manufacturing industries can be started and many older industries strengthened and expanded if we continue to study nature's laws and apply new knowledge to practical purposes.

Great advances in agriculture are also based upon scientific research. Plants which are more resistant to disease and are adapted to short growing seasons, the prevention and cure of livestock diseases, the control of our insect enemies, better fertilizers, and improved agricultural practices, all stem from painstaking scientific research.

Advances in science when put to practical use mean more jobs, higher wages, shorter hours, more abundant crops, more leisure for recreation, for study, for learning how to live without the deadening drudgery which has been the burden of the common man for ages past. Advances in science will also bring higher standards of living, will lead to the prevention or cure of diseases, will promote conservation of our limited national resources, and will assure means of defense against aggression. But to achieve these objectives—to secure a high level of employment, to maintain a position of world leadership—the flow of new scientific knowledge must be both continuous and substantial.

Our population increased from 75 million to 130 million between 1900 and 1940. In some countries comparable increases have been accompanied by famine. In this country the increase has been accompanied by more abundant food supply, better living, more leisure, longer life, and better health. This is, largely, the product of three factors—the free play of initiative of a vigorous people under democracy, the heritage of great natural wealth, and the advance of science and its application.

Science, by itself, provides no panacea for individual, social, and economic ills. It can be effective in the national welfare only as a member of a team, whether the conditions be peace or war. But without scientific progress no amount of achievement in other directions can insure our health, prosperity, and security as a nation in the modern world.

GOVERNMENT RESPONSIBILITY

It has been basic United States policy that Government should foster the opening of new frontiers. It opened the seas to clipper ships and furnished land for pioneers. Although these frontiers have more or less disappeared, the frontier of science remains. It is in keeping with the American tradition— one which has made the United States great—that new frontiers shall be made accessible for development by all American citizens.

Moreover, since health, well-being, and security are proper concerns of Government, scientific progress is, and must be, of vital interest to Government. Without scientific progress the national health would deteriorate; without scientific progress we could not hope for improvement in our standard of living or for an increased number of jobs for our citizens; and without scientific progress we could not have maintained our liberties against tyranny.

From early days the Government has taken an active interest in scientific matters. During the nineteenth century the Coast and Geodetic Survey, the Naval Observatory, the Department of Agriculture, and the Geological Survey were established. Through the Land Grant College Act the Government has supported research in state institutions for more than 80 years on a gradually increasing scale. Since 1900 a large number of scientific agencies have been established within the Federal Government, until in 1939 they numbered more than 40.

Much of the scientific research done by Government agencies is intermediate in character between the two types of work commonly referred to as basic and applied research. Almost all Government scientific work has ultimate practical objectives but, in many fields of broad national concern, it commonly involves long-term investigation of a fundamental nature. Generally speaking, the scientific agencies of Government are not so concerned with immediate practical objectives as are the laboratories of industry nor, on the other hand, are they so free to explore any natural phenomena without regard to possible economic applications as are the educational and private research institutions. Government scientific agencies have splendid records of achievement, but they are limited in function.

We have no national policy for science. The Government has only begun to utilize science in the nation's welfare. There is no body within the Government charged with formulating or executing a national science policy. There are no standing committees of the Congress devoted to this important subject. Science has been in the wings. It should be brought to the center of the stage—for in it lies much of our hope for the future.

There are areas of science in which the public interest is acute but which are likely to be cultivated inadequately if left without more support than will come from private sources. These areas—such as research on military problems, agriculture, housing, public health, certain medical research, and research involving expensive capital facilities beyond the capacity of private institutions—should be advanced by active Government support. To date, with the exception of the intensive war research conducted by the Office of Scientific Research and Development, such support has been meager and intermittent.

The publicly and privately supported colleges, universities, and research institutes are the centers of basic research. They are the wellsprings of knowledge and understanding. As long as they are vigorous and healthy and their scientists are free to pursue the truth wherever it may lead, there will be a flow of new scientific knowledge to those who can apply it to practical problems in Government, in industry, or elsewhere.

Many of the lessons learned in the war-time application of science under Government can be profitably applied in peace. The Government is peculiarly fitted to perform certain functions, such as the coordination and support of broad programs on problems of great national importance. But we must proceed with caution in carrying over the methods which work in wartime to the very different conditions of peace. We must remove the rigid controls which we have had to impose, and recover freedom of inquiry and that healthy competitive scientific spirit so necessary for expansion of the frontiers of scientic knowledge.

Scientific progress on a broad front results from the free play of free intellects, working on subjects of their own choice, in the manner dictated by their curiosity for exploration of the unknown. Freedom of inquiry must be preserved under any plan for Government support of science.

Note: Although Chapters 3 to 8 constitute a single unit, they have been separated for the convenience of the reader.

4: THE WAR AGAINST DISEASE

The death rate for all diseases in the Army, including the overseas forces, was reduced from 14.1 per thousand in the last war to 0.6 per thousand in World War II.

Such ravaging diseases as yellow fever, dysentery, typhus, tetanus, pneumonia, and meningitis have been all but conquered by penicillin and the sulfa drugs, the insecticide DDT, better vaccines, and improved hygienic measures. Malaria has been controlled. There has been dramatic progress in surgery.

The striking advances in medicine during the war were possible only because we had a large backlog of scientific data accumulated through basic research in many scientific fields in the years before the war.

In the last 40 years life expectancy in the United States has increased from 49 to 65 years largely as a consequence of the reduction in the death rates of infants and children; in the last 20 years the death rate from the diseases of childhood has been reduced 87 per cent.

Diabetes has been brought under control by insulin, pernicious anemia by liver extracts; and the once widespread deficiency diseases have been much reduced, even in the lowest income groups, by accessory food factors and improvement of diet. Notable advances have been made in the early diagnosis of cancer, and in the surgical and radiation treatment of the disease.

These results have been achieved through a great amount of basic research in medicine and the preclinical sciences, and by the dissemination of this new scientific knowledge through the physicians and medical services and public health agencies of the country. In this cooperative endeavor the pharmaceutical industry has played an important role, especially during the war. All of the medical and public health groups share credit for these achievements; they form interdependent members of a team.

Progress in combating disease depends upon an expanding body of new scientific knowledge.

UNSOLVED PROBLEMS

As President Roosevelt observed, the annual deaths from one or two diseases are far in excess of the total number of American lives lost in battle during this war. A large fraction of these deaths in our civilian population cut short the useful lives of our citizens. This is our present position despite the fact that in the last three decades notable progress has been made in civilian medicine. The reduction in death rate from diseases of childhood has shifted the emphasis to the middle and old age groups, particularly to the malignant diseases and the degenerative processes prominent in later life. Cardiovascular disease, including chronic disease of the kidneys, arteriosclerosis, and cerebral hemorrhage, now account for 45 per cent of the deaths in the United States. Second are the infectious diseases, and third is cancer. Added to these are many maladies (for example, the common cold, arthritis, asthma and hay fever, peptic ulcer) which, though infrequently fatal, cause incalculable disability.

Another aspect of the changing emphasis is the increase of mental diseases. Approximately 7 million persons in the United States are mentally ill; more than one-third of the hospital beds are occupied by such persons, at a cost of $175 million a year. Each year 125,000 new mental cases are hospitalized.

Notwithstanding great progress in prolonging the span of life and in relief of suffering, much illness remains for which adequate means of prevention and cure are not yet known. While additional physicians, hospitals, and health programs are needed, their full usefulness cannot be attained unless we enlarge our knowledge of the human organism and the nature of disease. Any extension of medical facilities must be accompanied by an expanded program of medical training and research.

Discoveries pertinent to medical progress have often come from remote and unexpected sources, and it is certain that this will be true in the future. It is wholly probable that progress in the treatment of cardiovascular disease, renal disease, cancer, and similar refractory diseases will be made as the result of fundamental discoveries in subjects unrelated to those diseases, and perhaps entirely unexpected by the investigator. Further progress requires that the entire front of medicine and the underlying sciences of chemistry, physics, anatomy, biochemistry, physiology, pharmacology, bacteriology, pathology, parasitology, etc., be broadly developed.

Progress in the war against disease results from discoveries in remote and unexpected fields of medicine and the underlying sciences.

Penicillin reached our troops in time to save countless lives because the Government coordinated and supported the program of research and development on the drug. The development moved from the early laboratory stage to large-scale production and use in a fraction of the time it would have taken without such leadership. The search for better anti-malarials, which proceeded at a moderate tempo for many years, has been accelerated enormously by Government support during the war. Other examples can be cited in which medical progress has been similarly advanced. In achieving these results, the Government has provided over-all coordination and support; it has not dictated how the work should be done within any cooperating institution.

Discovery of new therapeutic agents and methods usually results from basic studies in medicine and the underlying sciences. The development of such materials and methods to the point at which they become available to medical practitioners requires teamwork involving the medical schools, the science departments of universities, Government, and the pharmaceutical industry. Government initiative, support, and coordination can be very effective in this development phase.

Government initiative and support for the development of newly discovered therapeutic materials and methods can reduce the time required to bring the benefits to the public.

ACTION IS NECESSARY

The primary place for medical research is in the medical schools and universities. In some cases coordinated direct attack on special problems may be made by teams of investigators, supplementing similar attacks carried on by the Army, Navy, Public Health Service, and other organizations. Apart from teaching, however, the primary obligation of the medical schools and universities is to continue the traditional function of such institutions, namely, to provide the individual worker with an opportunity for free, untrammeled study of nature, in the directions and by the methods suggested by his interests, curiosity, and imagination. The history of medical science teaches clearly the supreme importance of affording the prepared mind complete freedom for the exercise of initiative. It is the special province of the medical schools and universities to foster medical research in this way—a duty which cannot be shifted to Government agencies, industrial organizations, or any other institutions.

Where clinical investigations of the human body are required, the medical schools are in a unique position, because of their close relationship to teaching hospitals, to integrate such investigations with the work of the departments of preclinical science, and to impart new knowledge to physicians in training. At the same time, the teaching hospitals are especially well qualified to carry on medical research because of their close connection with the medical schools, on which they depend for staff and supervision.

Between World War I and World War II the United States overtook all other nations in medical research and assumed a position of world leadership. To a considerable extent this progress reflected the liberal financial support from university endowment income, gifts from individuals, and foundation grants in the 20's. The growth of research departments in medical schools has been very uneven, however, and in consequence most of the important work has been done in a few large schools. This should be corrected by building up the weaker institutions, especially in regions which now have no strong medical research activities.

The traditional sources of support for medical research, largely endowment income, foundation grants, and private donations, are diminishing, and there is no immediate prospect of a change in this trend. Meanwhile, research costs have steadily risen. More elaborate and expensive equipment is required, supplies are more costly, and the wages of assistants are higher. Industry is only to a limited extent a source of funds for basic medical research.

It is clear that if we are to maintain the progress in medicine which has marked the last 25 years, the Government should extend financial support to basic medical research in the medical schools and in the universities, through grants both for research and for fellowships. The amount which can be effectively spent in the first year should not exceed 5 million dollars. After a program is under way perhaps 20 million dollars a year can be spent effectively.

5: THE PUBLIC WELFARE

In the late war it became clear beyond all doubt that scientific research is absolutely essential to national security. The bitter and dangerous battle against the U-boat was a battle of scientific techniques—and our margin of success was dangerously small. The new eyes which radar supplied to our fighting forces quickly evoked the development of scientific countermeasures which could often blind them. This again represents the ever continuing battle of techniques. The V-1 attack on London was finally defeated by three devices developed during the war and used superbly in the field. V-2 was countered only by capture of the launching sites.

The Secretaries of War and Navy recently stated in a joint letter to the National Academy of Sciences:

"This war emphasizes three facts of supreme importance to national security: (1) powerful new tactics of defense and offense are developed around new weapons created by scientific and engineering research; (2) the competitive time element in developing those weapons and tactics may be decisive; (3) war is increasingly total war, in which the armed services must be supplemented by active participation of every element of civilian population.

"To insure continued preparedness along farsighted technical lines, the research scientists of the country must be called upon to continue in peacetime some substantial portion of those types of contribution to national security which they have made so effectively during the stress of the present war. . . ."

There must be more—and more adequate—military research during peacetime. We cannot again rely on our allies to hold off the enemy while we struggle to catch up. Further, it is clear that only the Government can undertake military research; for it must be carried on in secret, much of it has no commercial value, and it is expensive. The obligation of Government to support research on military problems is inescapable.

Modern war requires the use of the most advanced scientific techniques. Many of the leaders in the development of radar are scientists who before the war had been exploring the nucleus of the atom. While there must be increased emphasis on science in the future training of officers for both the Army and Navy, such men cannot be expected to be specialists in scientific research. Therefore a professional partnership between the officers in the Services and civilian scientists is needed.

The Army and Navy should continue to carry on research and development in the improvement of current weapons. For many years the National Advisory Committee for Aeronautics has supplemented the work of the Army and Navy by conducting basic research on the problems of flight. There should now be permanent civilian activity to supplement the research work of the Services in other scientific fields so as to carry on in time of peace some part of the activities of the emergency war-time Office of Scientific Research and Development.

Military preparedness requires a permanent independent, civilian-controlled organization, having close liaison with the Army and Navy, but with funds directly from Congress and with the clear power to initiate military research which will supplement and strengthen that carried on directly under the control of the Army and Navy.

One of our hopes is that there will be full employment, and that the production of goods and services will serve to raise our standard of living. We do not know yet how we shall reach that goal, but it is certain that it can be achieved only by releasing the full creative and productive energies of the American people.

Surely we shall not get there by standing still, merely by making the same things we made before and selling them at the same or higher prices. We shall not get ahead in international trade unless we offer new and more attractive and cheaper products.

Where will these new products come from? How shall we find ways to make better products at lower cost? The answer is clear. There must be a stream of new scientific knowledge to turn the wheels of private and public enterprise. There must be plenty of men and women trained in science and technology, for upon them depend both the creation of new knowledge and its application to practical purposes.

More and better scientific research is essential to the achievement of our goal of full employment.

THE IMPORTANCE OF BASIC RESEARCH

Basic research is performed without thought of practical ends. It results in general knowledge and an understanding of nature and its laws. This general knowledge provides the means of answering a large number of important practical problems, though it may not give a complete specific answer to any one of them. The function of applied research is to provide such complete answers. The scientist doing basic research may not be at all interested in the practical applications of his work, yet the further progress of industrial development would eventually stagnate if basic scientific research were long neglected.

One of the peculiarities of basic science is the variety of paths which lead to productive advance. Many of the most important discoveries have come as a result of experiments undertaken with very different purposes in mind. Statistically it is certain that important and highly useful discoveries will result from some fraction of the undertakings in basic science; but the results of any one particular investigation cannot be predicted with accuracy.

Basic research leads to new knowledge. It provides scientific capital. It creates the fund from which the practical applications of knowledge must be drawn. New products and new processes do not appear full-grown. They are founded on new principles and new conceptions, which in turn are painstakingly developed by research in the purest realms of science.

Today, it is truer than ever that basic research is the pacemaker of technological progress. In the nineteenth century, Yankee mechanical ingenuity, building largely upon the basic discoveries of European scientists, could greatly advance the technical arts. Now the situation is different.

A nation which depends upon others for its new basic scientific knowledge will be slow in its industrial progress and weak in its competitive position in world trade, regardless of its mechanical skill.

Publicly and privately supported colleges and universities and the endowed research institutes must furnish both the new scientific knowledge and the trained research workers. These institutions are uniquely qualified by tradition and by their special characteristics to carry on basic research. They are charged with the responsibility of conserving the knowledge accumulated by the past, imparting that knowledge to students, and contributing new knowledge of all kinds. It is chiefly in these institutions that scientists may work in an atmosphere which is relatively free from the adverse pressure of convention, prejudice, or commercial necessity. At their best they provide the scientific worker with a strong sense of solidarity and security, as well as a substantial degree of personal intellectual freedom. All of these factors are of great importance in the development of new knowledge, since much of new knowledge is certain to arouse opposition because of its tendency to challenge current beliefs or practice.

Industry is generally inhibited by preconceived goals, by its own clearly defined standards, and by the constant pressure of commercial necessity. Satisfactory progress in basic science seldom occurs under conditions prevailing in the normal industrial laboratory. There are some notable exceptions, it is true, but even in such cases it is rarely possible to match the universities in respect to the freedom which is so important to scientific discovery.

To serve effectively as the centers of basic research these institutions must be strong and healthy They must attract our best scientists as teachers and investigators. They must offer research opportunities and sufficient compensation to enable them to compete with industry and government for the cream of scientific talent.

During the past 25 years there has been a great increase in industrial research involving the application of scientific knowledge to a multitude of practical purposes—thus providing new products, new industries, new investment opportunities, and millions of jobs. During the same period research within Government—again largely applied research—has also been greatly expanded. In the decade from 1930 to 1940 expenditures for industrial research increased from $116,000,000 to $240,000,-000 and those for scientific research in Government rose from $24,000,000 to $69,000,000. During the same period expenditures for scientific research in the colleges and universities increased from $2,000,000 to $31,000,000, while those in the endowed research institutes declined from $5,200,000 to $4,500,000. These are the best estimates available. The figures have been taken from a variety of sources and arbitrary definitions have necessarily been applied, but it is believed that they may be accepted as indicating the following trends:

(a) Expenditures for scientific research by industry and Government—almost entirely applied research—have more than doubled between 1930 and 1940. Whereas in 1930 they were six times as large as the research expenditures of the colleges, universities, and research institutes, by 1940 they were nearly ten times as large.

(b) While expenditures for scientific research in the colleges and universities increased by one-half during this period, those for the endowed research institutes have slowly declined.

If the colleges, universities, and research institutes are to meet the rapidly increasing demands of industry and Government for new scientific knowledge, their basic research should be strengthened by use of public funds.

RESEARCH WITHIN THE GOVERNMENT

Although there are some notable exceptions, most research conducted within Governmental laboratories is of an applied nature. This has always been true and is likely to remain so. Hence Government, like industry, is dependent upon the colleges, universities, and research institutes to expand the basic scientific frontiers and to furnish trained scientific investigators.

Research within the Government represents an important part of our total research activity and needs to be strengthened and expanded. Such expansion should be directed to fields of inquiry and service which are of public importance and are not adequately carried on by private organizations.

The most important single factor in scientific and technical work is the quality of personnel employed. The procedures currently followed within the Government for recruiting, classifying and compensating such personnel place the Government under a severe handicap in competing with industry and the universities for first-class scientific talent. Steps should be taken to reduce that handicap.

In the Government the arrangement whereby the numerous scientific agencies form parts of larger departments has both advantages and disadvantages. But the present pattern is firmly established and there is much to be said for it. There is, however, a very real need for some measure of coordination of the common scientific activities of these agencies, as to both policies and budgets, and at present no such means exists.

A permanent Science Advisory Board should be created to consult with these scientific bureaus and to advise the executive and legislative branches of Government as to the policies and budgets of Government agencies engaged in scientific research. This board should be composed of disinterested scientists who have no connection with the affairs of any Government agency.

INDUSTRIAL RESEARCH

The simplest and most effective way in which the Government can strengthen industrial research is to support basic research and to develop scientific talent.

The benefits of basic research do not reach all industries equally or at the same speed. Some small enterprises never receive any of the benefits. It has been suggested that the benefits might be better utilized if "research clinics" for such enterprises were to be established. Businessmen would thus be able to make more use of research than they now do. This proposal is certainly worthy of further study.

One of the most important factors affecting the amount of industrial research is the income-tax law. Government action in respect to this subject will affect the rate of technical progress in industry. Uncertainties as to the attitude of the Bureau of Internal Revenue regarding the deduction of research and development expenses are a deterrent to research expenditure. These uncertainties arise from lack of clarity of the tax law as to the proper treatment of such costs. The Internal Revenue Code should be amended to remove present uncertainties in regard to the deductibility of research and development expenditures as current charges against net income.

Research is also affected by the patent laws. They stimulate new invention and they make it possible for new industries to be built around new devices or new processes. These industries generate new jobs and new products, all of which contribute to the welfare and the strength of the country.

Yet, uncertainties in the operation of the patent laws have impaired the ability of small industries to translate new ideas into processes and products of value to the nation. These uncertainties are, in part, attributable to the difficulties and expense incident to the operation of the patent system as it presently exists. These uncertainties are also attributable to the existence of certain abuses which have appeared in the use of patents. The abuses should be corrected. They have led to extravagantly critical attacks which tend to discredit a basically sound system.

It is important that the patent system continue to serve the country in the manner intended by the Constitution, for it has been a vital element in the industrial vigor which has distinguished this nation.

The National Patent Planning Commission has reported on this subject. In addition, a detailed study, with recommendations concerning the extent to which modifications should be made in our patent laws, is currently being made under the leadership of the Secretary of Commerce. It is recommended, therefore, that specific action with regard to the patent laws be withheld pending the submission of the report devoted exclusively to that subject.

INTERNATIONAL EXCHANGE OF INFORMATION

International exchange of scientific information is of growing importance. Increasing specialization of science will make it more important than ever that scientists in this country keep continually abreast of developments abroad. In addition a flow of scientific information constitutes one facet of general international accord which should be cultivated.

The Government can accomplish significant results in several ways: by aiding in the arrangement of international science congresses, in the official accrediting of American scientists to such gatherings, in the official reception in this country of foreign scientists of standing, in making possible a rapid flow of technical information, including translation service, and possibly in the provison of international fellowships. Private foundations and other groups partially fulfill some of these functions at present, but their scope is incomplete and inadequate.

The Government should take an active role in promoting the international flow of scientific information.

THE SPECIAL NEED FOR FEDERAL SUPPORT

We can no longer count on ravaged Europe as a source of fundamental knowledge. In the past we have devoted much of our best efforts to the application of such knowledge which has been discovered abroad. In the future we must pay increased attention to discovering this knowledge for ourselves, particularly since the scientific applications of the future will be more than ever dependent upon such basic knowledge.

New impetus must be given to research in our country. Such new impetus can come promptly only from the Government. Expenditures for research in the colleges, universities, and research institutes will otherwise not be able to meet the additional demands of increased public need for research.

Further, we cannot expect industry adequately to fill the gap. Industry will fully rise to the challenge of applying new knowledge to new products. The commercial incentive can be relied upon for that. But basic research is essentially noncommercial in nature. It will not receive the attention it requires if left to industry.

For many years the Government has wisely supported research in the agricultural colleges and the benefits have been great. The time has come when such support should be extended to other fields.

In providing Government support, however, we must endeavor to preserve as far as possible the private support of research both in industry and in the colleges, universities, and research institutes. These private sources should continue to carry their share of the financial burden.

It is estimated that an adequate program for Federal support of basic research in the colleges, universities, and research institutes and for financing important applied research in the public interest, will cost about 10 million dollars at the outset and may rise to about 50 million dollars annually when fully under way at the end of perhaps 5 years.

6: RENEWAL OF SCIENTIFIC TALENT

The responsibility for the creation of new scientific knowledge rests on that small body of men and women who understand the fundamental laws of nature and are skilled in the techniques of scientific research. While there will always be the rare individual who will rise to the top without benefit of formal education and training, he is the exception and even he might make a more notable contribution if he had the benefit of the best education we have to offer. I cannot improve on President Conant's statement that:

". . . in every section of the entire area where the word science may properly be applied, the limiting factor is a human one. We shall have rapid or slow advance in this direction or in that depending on the number of really first-class men who are engaged in the work in question. . . . So in the last analysis, the future of science in this country will be determined by our basic educational policy."

It would be folly to set up a program under which research in the natural sciences and medicine was expanded at the cost of the social sciences, humanities, and other studies so essential to national well-being. This point has been well stated by the Moe Committee as follows:

"As citizens, as good citizens, we therefore think that we must have in mind while examining the question before us— the discovery and development of scientific talent—the needs of the whole national welfare. We could not suggest to you a program which would syphon into science and technology a disproportionately large share of the nation's highest abilities, without doing harm to the nation, nor, indeed, without crippling science. . . . Science cannot live by and unto itself alone. . . .

"The uses to which high ability in youth can be put are various and, to a large extent, are determined by social pressures and rewards. When aided by selective devices for picking out scientifically talented youth, it is clear that large sums of money for scholarships and fellowships and monetary and other rewards in disproportionate amounts might draw into science too large a percentage of the nation's high ability, with a result highly detrimental to the nation and to science. Plans for the discovery and development of scientific talent must be related to the other needs of society for high ability. . . . There is never enough ability at high levels to satisfy all the needs of the nation; we would not seek to draw into science any more of it than science's proportionate share."

THE WARTIME DEFICIT

Among the young men and women who are qualified to take up scientific work, few over 18 have been found since 1940 following an integrated course of scientific studies, except for some in medicine and engineering in Army and Navy programs and a few 4-F's. Neither our Allies nor, so far as we know, our enemies have done anything so radical as thus to suspend almost completely their educational activities in scientific pursuits during the war period.

Two great principles guided us in this country as we turned our full efforts to war: First, the sound democratic principle that there should be no favored classes or special privilege in a time of peril, that all should be ready to sacrific equally; second, the tenet that every man should serve in the capacity in which his talents and experience can best be applied for the prosecution of the war effort. In general we have held these principles well in balance.

In my opinion, however, we drew too heavily for non-scientific purposes upon the great natural resource which resides in our trained young scientists and engineers. For the general good of the country too many such men went into uniform, and their talents were not always fully utilized. With the exception of those men engaged in war research, all physically fit students at graduate level were taken into the armed forces. Those ready for college training in the sciences were not permitted to enter upon that training.

As a result an accumulating deficit of trained research personnel will continue for many years. The deficit of science and technology students who, but for the war, would have received bachelor's degrees is about 150,000. The deficit of those holding advanced degrees—that is, young scholars trained to the point where they are capable of carrying on original work—has been estimated as amounting to about 17,000 by 1955 in chemistry, engineering, geology, mathematics, physics, psychology, and the biological sciences.

With mounting demands for scientists both for teaching and for research, we entered the postwar period with a serious deficit in our trained scientific personnel.

NEED FOR A PROGRAM

Confronted with these deficits, we are compelled to look to the use of our basic human resources and formulate a program which will assure their conservation and effective development. The committee advising me on scientific personnel has stated the following principle which should guide our planning:

"If we were all-knowing and all-wise we might, but we think probably not, write you a plan whereby there might be selected for training, which they otherwise would not get, those who, 20 years hence, would be scientific leaders, and we might not bother about any lesser manifestations of scientific ability. But in the present state of knowledge a plan cannot be made which will select, and assist, only those young men and women who will give the top future leadership to science. To get top leadership there must be a relatively large base of high ability selected for development and then successive skimmings of the cream of ability at successive times and at higher levels. No one can select from the bottom those who will be the leaders at the top because unmeasured and unknown factors enter into scientific, or any, leadership. There are brains and character, strength and health, happiness and spiritual vitality, interest and motivation, and no one knows what else, that must needs enter into this supra-mathematical calculus.

"We think we probably would not, even if we were all-wise and all-knowing, write you a plan whereby you would be assured of scientific leadership at one stroke. We think as we think because we are not interested in setting up an elect. We think it much the best plan, in this constitutional Republic, that opportunity be held out to all kinds and conditions of men whereby they can better themselves. This is the American way; this is the way the United States has become what it is. We think it very important that circumstances be such that there be no ceilings, other than ability itself, to intellectual ambition. We think it very important that every boy and girl shall know that, if he shows that he has what it takes, the sky is the limit. Even if it be shown subsequently that he has not what it takes to go to the top, he will go further than he would otherwise go if there had been a ceiling beyond which he always knew he could not aspire.

"By proceeding from point to point and taking stock on the way, by giving further opportunity to those who show themselves worthy of further opportunity, by giving the most opportunity to those who show themselves continually developing —this is the way we propose. This is the American way: a man works for what he gets."

Higher education in this country is largely for those who have the means. If those who have the means coincided entirely with those persons who have the talent we should not be squandering a part of our higher education on those undeserving of it, nor neglecting great talent among those who fail to attend college for economic reasons. There are talented individuals in every segment of the population, but with few exceptions those without the means of buying higher education go without it. Here is a tremendous waste of the greatest resource of a nation—the intelligence of its citizens.

If ability, and not the circumstance of family fortune, is made to determine who shall receive higher education in science, then we shall be assured of constantly improving quality at every level of scientific activity.

We have a serious deficit in scientific personnel partly because the men who would have studied science in the colleges and universities have been serving in the Armed Forces. Many had begun their studies before they went to war. Others with capacity for scientific education went to war after finishing high school. The most immediate prospect of making up some of the deficit in scientific personnel is by salvaging scientific talent from the generation in uniform. For even if we should start now to train the current crop of high school graduates, it would be 1951 before they would complete graduate studies and be prepared for effective scientific research. This fact underlines the necessity of salvaging potential scientists in uniform.

The Armed Services should comb their records for men who, prior to or during the war, have given evidence of talent for science, and make prompt arrangements, consistent with current discharge plans, for ordering those who remain in uniform as soon as militarily possible to duty at institutions here and overseas where they can continue their scientific education. Moreover, they should see that those who study overseas have the benefit of the latest scientific developments.

The country may be proud of the fact that 95 per cent of boys and girls of fifth grade age are enrolled in school, but the drop in enrollment after the fifth grade is less satisfying. For every 1,000 students in the fifth grade, 600 are lost to education before the end of high school, and all but 72 have ceased formal education before completion of college. While we are concerned primarily with methods of selecting and educating high school graduates at the college and higher levels, we cannot be complacent about the loss of potential talent which is inherent in the present situation.

Students drop out of school, college, and graduate school, or do not get that far, for a variety of reasons: they cannot afford to go on; schools and colleges providing courses equal to their capacity are not available locally; business and industry recruit many of the most promising before they have finished the training of which they are capable. These reasons apply with particular force to science: the road is long and expensive; it extends at least 6 years beyond high school; the percentage of science students who can obtain first-rate training in institutions near home is small.

Improvement in the teaching of science is imperative, for students of latent scientific ability are particularly vulnerable to high school teaching which fails to awaken interest or to provide adequate instruction. To enlarge the group of specially qualified men and women it is necessary to increase the number who go to college. This involves improved high school instruction, provision for helping individual talented students to finish high school (primarily the responsibility of the local communities), and opportunities for more capable, promising high school students to go to college. Anything short of this means serious waste of higher education and neglect of human resources.

To encourage and enable a larger number of young men and women of ability to take up science as a career, and in order gradually to reduce the deficit of trained scientific personnel, it is recommended that provision be made for a reasonable number of (a) undergraduate scholarships and graduate fellowships and (b) fellowships for advanced training and fundamental research. The details should be worked out with reference to the interests of the several States and of the universities and colleges, and care should be taken not to impair the freedom of the institutions and individuals concerned.

The program proposed by the Moe Committee would provide 24,000 undergraduate scholarships and 900 graduate fellowships and would cost about $30,000,000 annually when in full operation. Each year under this program 6,000 undergraduate scholarships would be made available to high school graduates, and 300 graduate fellowships would be offered to college graduates. Approximately the scale of allowances provided for under the educational program for returning veterans has been used in estimating the cost of this program.

The plan is, further, that all those who receive such scholarships or fellowships in science should be enrolled in a National Science Reserve and be liable to call into the service of the Government, in connection with scientific or technical work in time of war or other national emergency declared by Congress or proclaimed by the President. Thus, in addition to the general benefits to the nation by reason of the addition to its trained ranks of such a corps of scientific workers, there would be a definite benefit to the nation in having these scientific workers on call in national emergencies. The Government would be well advised to invest the money involved in this plan even if the benefits to the nation were thought of solely—as they are not—in terms of national preparedness.

7: RECONVERSION OPPORTUNITIES

We have been living on our fat. For more than five years many of our scientists fought the war in the laboratories, in the factories and shops, and at the front. We directed the energies of our scientists to the development of weapons and materials and methods, on a large number of relatively narrow projects initiated and controlled by the Office of Scientific Research and Development and other Government agencies. Like troops, the scientists were mobilized and thrown into action to serve their country in time of emergency. But they were diverted to a greater extent than is generally appreciated from the search for answers to the fundamental problems—from the search on which human welfare and progress depend. This is not a complaint—it is a fact. The mobilization of science behind the lines aided the fighting men at the front to win the war and to shorten it; and it resulted incidentally in the accumulation of a vast amount of experience and knowledge of the application of science to particular problems, much of which can be put to use now that the war is over. Fortunately, this country had the scientists—and the time—to make this contribution and thus to advance the date of victory.

Much of the information and experience acquired during the war is confined to the agencies that gathered it. Except to the extent that military security dictates otherwise, such knowledge should be spread upon the record for the benefit of the public.

Thanks to the wise provision of the Secretary of War and the Secretary of the Navy, most of the results of wartime medical research have been published. The material still subject to security classification should be released as soon as possible.

It is my view that most of the remainder of the classified scientific material should be released as soon as is practicable. Most of the information needed by industry and in education can be released without disclosing its embodiments in actual military material and devices. Basically there is no reason to believe that scientists of other countries will not in time rediscover everything we now know which is held in secrecy. A broad dissemination of scientific information upon which further advances can readily be made furnishes a sounder foundation for our national security than a policy of restriction which would impede our own progress although imposed in the hope that possible enemies would not catch up with us.

During the war it was necessary for selected groups of scientists to work on specialized problems, with relatively little information as to what other groups were doing and had done. Working against time, the Office of Scientific Research and Development was obliged to enforce this practice during the war, although it was realized by all concerned that it was an emergency measure which prevented the continuous cross-fertilization so essential to fruitful scientific effort.

Our ability to overcome possible future enemies depends upon scientific advances, which will proceed more rapidly with diffusion of knowledge than under a policy of continued restriction of knowledge now in our possession.

COORDINATION AND DISSEMINATION

In planning the release of scientific data and experience collected in connection with the war, we must not overlook the fact that research has gone forward under many auspices—the Army, the Navy, the Office of Scientific Research and Development, the National Advisory Committee for Aeronautics, other departments and agencies of the Government, educational institutions, and many industrial organizations. There have been numerous cases of independent discovery of the same truth in different places. To permit the release of information by one agency and to continue to restrict it elsewhere would be unfair in its effect and would tend to impair the morale and efficiency of scientists who submerged individual interests in the controls and restrictions of war.

A part of the information now classified which should be released is possessed jointly by our allies and ourselves. Plans for release of such information should be coordinated with our allies to minimize danger of international friction which would result from sporadic uncontrolled release.

The agency responsible for recommending the release of information from military classification should be an Army, Navy, civilian body, well grounded in science and technology. It should be competent to advise the Secretary of War and the Secretary of the Navy. It should, moreover, have sufficient recognition to secure prompt and practical decisions. To satisfy these considerations I recommend the establishment of a Board, made up equally of scientists and military men, whose function would be to pass upon the declassification and to control the release for publication of scientific information which is now classified.

The release of information from security regulations is but one phase of the problem. The other is to provide for preparation of the material and its publication in a form and at a price which will facilitate dissemination and use. In the case of the Office of Scientific Research and Development, arrangements have been made for the preparation of manuscripts while the staffs are still assembled and in possession of the records.

We should get this scientific material to scientists everywhere with great promptness, and at as low a price as is consistent with suitable format. We should also get it to the men studying overseas so that they will know what has happened in their absence.

It is recommended that measures which will encourage and facilitate the preparation and publication of reports be adopted forthwith by all agencies, governmental and private, possessing scientific information released from security control.

8: THE MEANS TO THE END

The Federal Government should accept new responsibilities for promoting the creation of new scientific knowledge and the development of scientific talent in our youth. In discharging these responsibilities, Federal funds should be made available. We have given much thought to the question of how plans for the use of Federal funds may be arranged so that such funds will not drive out of the picture funds from local governments, foundations, and private donors. We believe that our proposals will minimize that effect, but we do not think that it can be completely avoided. We submit, however, that the nation's need for more and better scientific research is such that the risk must be accepted.

It is also clear that the effective discharge of these responsibilities will require the full attention of some over-all agency devoted to that purpose. There should be a focal point within the Government for a concerted program of assisting scientific research conducted outside of Government. Such an agency should furnish the funds needed to support basic research in the colleges and universities, should coordinate where possible research programs on matters of utmost importance to the national welfare, should formulate a national policy for the Government toward science, should sponsor the interchange of scientific information among scientists and laboratories both in this country and abroad, and should ensure that the incentives to research in industry and the universities are maintained.

There are within Government departments many groups whose interests are primarily those of scientific research. Notable examples are found within the Departments of Agriculture, Commerce, Interior, and the Federal Security Agency.

These groups are concerned with science as collateral and peripheral to the major problems of those departments. These groups should remain where they are, and continue to perform their present functions, including the support of agricultural research by grants to the Land Grant Colleges and Experiment Stations, since their largest contribution lies in applying fundamental knowledge to the special problems of the departments within which they are established.

By the same token these groups cannot be made the repository of the new and large responsibilities in science which belong to the Government and which the Government should accept. The recommendations which relate to research within the Government, to the release of scientific information, to clarification of the tax laws, and to the recovery and development of our scientific talent now in uniform can be implemented by action within the existing structure of the Government. But nowhere in the Governmental structure receiving its funds from Congress is there an agency adapted to supplementing the support of basic research in the universities, in both medicine and the natural sciences; adapted to supporting research on new weapons for both Services; or adapted to administering a program of science scholarships and fellowships.

A new agency should be established, therefore, by the Congress for the purpose. Such an agency, moreover, should be an independent agency devoted to the support of scientific research and advanced scientific education alone. Industry learned many years ago that basic research cannot often be fruitfully conducted as an adjunct to or a subdivision of an operating agency or department. Operating agencies have immediate operating goals and are under constant pressure to produce in a tangible way, for that is the test of their value. None of these conditions is favorable to basic research. Research is the exploration of the unknown and is necessarily speculative. It is inhibited by conventional approaches, traditions, and standards. It cannot be satisfactorily conducted in an atmosphere where it is gauged and tested by operating or production standards. Basic scientific research should not, therefore, be placed under an operating agency whose paramount concern is anything other than research. Research will always suffer when put in competition with operations.

I am convinced that these new functions should be centered in one agency. Science is fundamentally a unitary thing. The number of independent agencies should be kept to a minimum. Much medical progress, for example, will come from fundamental advances in chemistry. Separation of the sciences in tight compartments, as would occur if more than one agency were involved, would retard and not advance scientific knowledge as a whole.

FIVE FUNDAMENTALS

There are certain basic principles which must underlie the program of Government support for scientific research and education if such support is to be effective and if it is to avoid impairing the very things we seek to foster. These principles are as follows:

(1) Whatever the extent of support may be, there must be stability of funds over a period of years so that long-range programs may be undertaken.

(2) The agency to administer such funds should be composed of citizens selected only on the basis of their interest in and capacity to promote the work of the agency. They should be persons of broad interest in and understanding of the peculiarities of scientific research and education.

(3) The agency should promote research through contracts or grants to organizations outside the Federal Government. It should not operate any laboratories of its own.

(4) Support of basic research in the public and private colleges, universities, and research institutes must leave the internal control of policy, personnel, and the method and scope of the research to the institutions themselves. This is of the utmost importance.

(5) While assuring complete independence and freedom for

the nature, scope, and methodology of research carried on in the institutions receiving public funds, and while retaining discretion in the allocation of funds among such institutions, the Foundation proposed herein must be responsible to the President and the Congress. Only through such responsibility can we maintain the proper relationship between science and other aspects of a democratic system. The usual controls of audits, reports, budgeting, and the like should, of course, apply to the administrative and fiscal operations of the Foundation, subject, however, to such adjustments in procedure as are necessary to meet the special requirements of research.

Basic research is a long-term process—it ceases to be basic if immediate results are expected on short-term support. Methods should therefore be found which will permit the agency to make commitments of funds from current appropriations for programs of five years' duration or longer. Continuity and stability of the program and its support may be expected (*a*) from the growing realization by the Congress of the benefits to the public from scientific research, and (*b*) from the conviction which will grow among those who conduct research under the auspices of the agency that good quality work will be followed by continuing support.

MILITARY RESEARCH

As stated earlier, military preparedness requires a permanent, independent, civilian-controlled organization, having close liaison with the Army and Navy, but with funds direct from Congress and the clear power to initiate military research which will supplement and strengthen that carried on directly under the control of the Army and Navy. As a temporary measure the National Academy of Sciences established the Research Board for National Security at the request of the Secretary of War and the Secretary of the Navy in order to avert interruption in the relations between scientists and military men after the termination of emergency wartime organization.

I believe that, as a permanent measure, it would be appropriate to add to the agency needed to perform the other functions recommended the responsibilities for civilian-initiated and civilian-controlled military research. The function of such a civilian group would be primarily to conduct long-range scientific research on military problems—leaving to the Services research on the improvement of existing weapons.

Some research on military problems should be conducted, in time of peace as well as in war, by civilians independently of the military establishment. It is the primary responsibility of the Army and Navy to train the men, make available the weapons, and employ the strategy that will bring victory in combat. The Armed Services cannot be expected to be experts in all of the complicated fields which make it possible for a great nation to fight successfully in total war. There are certain kinds of research—such as research on the improvement of existing weapons—which can best be done within the military establishment. However, the job of long-range research involving application of the newest scientific discoveries to military needs should be the responsibility of those civilian scientists in the universities and in industry who are best trained to discharge it thoroughly and successfully. It is essential that both kinds of research go forward and that there be the closest liaison between the two groups.

Placing the civilian military research function in the proposed agency would bring it into close relationship with a broad program of basic research in both the natural sciences and medicine. A balance between military and other research could thus readily be maintained.

The establishment of the new agency, including a civilian military research group, should not be delayed by the existence of the Research Board for National Security, which is a temporary measure. Nor should the creation of the new agency be delayed by uncertainties in regard to the postwar organization of our military departments themselves. Clearly, the new agency, including a civilian military research group within it, can remain sufficiently flexible to adapt its operations to whatever may be the final organization of the military departments.

NATIONAL RESEARCH FOUNDATION

It is my judgment that the national interest in scientific research and scientific education can best be promoted by the creation of a National Research Foundation.

Purposes: The National Research Foundation should develop and promote a national policy for scientific research and scientific education, should support basic research in nonprofit organizations, should develop scientific talent in American youth by means of scholarships and fellowships, and should, by contract and otherwise, support long-range research on military matters.

Members: Responsibility to the people, through the President and the Congress, should be placed in the hands of, say, nine Members, who should be persons not otherwise connected with the Government and not representative of any special interest, who should be known as National Research Foundation Members, selected by the President on the basis of their interest in and capacity to promote the purposes of the Foundation.

The terms of the Members should be, say, 4 years, and no Member should be eligible for immediate reappointment provided he has served a full 4-year term. It should be arranged that the Members first appointed serve terms of such length that at least two Members are appointed each succeeding year.

The Members should serve without compensation but should be entitled to their expenses incurred in the performance of their duties.

The Members should elect their own chairman annually.

The chief executive officer of the Foundation should be a director appointed by the Members. Subject to the direction and supervision of the Foundation Members (acting as a board), the director should discharge all the fiscal, legal, and administrative functions of the Foundation. The director should receive a salary that is fully adequate to attract an outstanding man to the post.

There should be an administrative office responsible to the director to handle in one place the fiscal, legal, personnel, and other similar administrative functions necessary to the accomplishment of the purposes of the Foundation.

With the exception of the director, the division members, and one executive officer appointed by the director to administer the affairs of each division, all employees of the Foundation should be appointed under Civil Service regulations.

Organization: In order to accomplish the purposes of the Foundation the Members should establish several professional Divisions to be responsible to the Members. At the outset these Divisions should be:

Division of Medical Research; the function of this Division should be to support medical research.

Division of Natural Sciences; the function of this Division should be to support research in the physical and natural sciences.

Division of National Defense; it should be the function of this Division to support long-range scientific research on military matters.

Division of Scientific Personnel and Education; it should be the function of this Division to support and to supervise the grant of scholarships and fellowships in science.

Division of Publications and Scientific Collaboration; this Division should be charged with encouraging the publication of scientific knowledge and promoting international exchange of scientific information.

Each Division of the Foundation should be made up of at least five members, appointed by the Members of the Foundation. In making such appointments the Members should request and consider recommendations from the National Academy of Sciences which should be asked to establish a new National Research Foundation nominating committee in order to bring together the recommendations of scientists in all organizations. The chairman of each Division should be appointed by the Members of the Foundation.

The Division Members should be appointed for such terms as the Members of the Foundation may determine, and may be reappointed at the discretion of the Members. They should receive their expenses and compensation for their services at a per diem rate of, say, $50 while engaged on business of the Foundation, but no division member should receive more than, say, $10,000 compensation per year.

Membership of the Division of National Defense should include in addition to, say, five civilian members, one representative designated by the Secretary of War, and one representative designated by the Secretary of the Navy, who should serve without additional compensation for this duty.

Functions: The Members of the Foundation should have the following functions, powers, and duties:

To formulate over-all policies of the Foundation.

To establish and maintain such offices within the United States, its territories and possessions, as they may deem necessary.

To meet and function at any place within the United States, its territories and possessions.

To obtain and utilize the services of other Government agencies to the extent that such agencies are prepared to render such services.

To adopt, promulgate, amend, and rescind rules and regulations to carry out the provisions of the legislation and the policies and practices of the Foundation.

To review and balance the financial requirements of the several Divisions and to propose to the President the annual estimate for the funds required by each Division. Appropriations should be earmarked for the purposes of specific Divisions, but the Foundation should be left discretion with respect to the expenditure of each Division's funds.

To make contracts or grants for the conduct of research by negotiation without advertising for bids.

And with the advice of the National Research Foundation Divisions concerned—

To create such advisory and cooperating agencies and councils, state, regional, or national, as in their judgment will aid in effectuating the purposes of the legislation, and to pay the expenses thereof.

To enter into contracts with or make grants to educational and nonprofit research institutions for support of scientific research.

To initiate and finance in appropriate agencies, institutions, or organizations, research on problems related to the national defense.

To initiate and finance in appropriate organizations research projects for which existing facilities are unavailable or inadequate.

To establish scholarships and fellowships in the natural sciences including biology and medicine.

To promote the dissemination of scientific and technical information and to further its international exchange.

To support international cooperation in science by providing financial aid for international meetings, associations of scientific societies, and scientific research programs organized on an international basis.

To devise and promote the use of methods of improving the transition between research and its practical application in industry.

The Divisions should be responsible to the Members of the Foundation for—

Formulation of programs and policy within the scope of the particular Divisions.

Recommendation regarding the allocation of research programs among research organizations.

Recommendation of appropriate arrangements between

the Foundation and the organizations selected to carry on the program.

Recommendation of arrangements with State and local authorities in regard to cooperation in a program of science scholarships and fellowships.

Periodic review of the quality of research being conducted under the auspices of the particular Division and revision of the program of support of research.

Presentation of budgets of financial needs for the work of the Division.

Maintaining liaison with other scientific research agencies, both governmental and private, concerned with the work of the Division.

Patent Policy: The success of the National Research Foundation in promoting scientific research in this country will depend to a very large degree upon the cooperation of organizations outside the Government. In making contracts with or grants to such organizations the Foundation should protect the public interest adequately and at the same time leave the co-operating organization with adequate freedom and incentive to conduct scientific research. The public interest will normally be adequately protected if the Government receives a royalty-free license for governmental purposes under any patents resulting from work financed by the Foundation. There should be no obligation on the research institution to patent discoveries made as a result of support from the Foundation. There should certainly *not* be any absolute requirement that all rights in such discoveries be assigned to the Government, but it should be left to the discretion of the director and the interested Division whether in special cases the public interest requires such an assignment. Legislation on this point should leave to the Members of the Foundation discretion as to its patent policy in order that patent arrangements may be adjusted as circumstances and the public interest require.

Special Authority: In order to insure that men of great competence and experience may be designated as Members of the

Foundation and as members of the several professional Divisions, the legislation creating the Foundation should contain specific authorization so that the Members of the Foundation and the Members of the Divisions may also engage in private and gainful employment, notwithstanding the provisions of any other laws: provided, however, that no compensation for such employment is received in any form from any profit-making institution which receives funds under contract, or otherwise, from the Division or Divisions of the Foundation with which the individual is concerned. In normal times, in view of the restrictive statutory prohibitions against dual interests on the part of Government officials, it would be virtually impossible to persuade persons having private employment of any kind to serve the Government in an official capacity. In order, however, to secure the part-time services of the most competent men as Members of the Foundation and the Divisions, these stringent prohibitions should be relaxed to the extent indicated.

Since research is unlike the procurement of standardized items, which are susceptible to competitive bidding on fixed specifications, the legislation creating the National Research Foundation should free the Foundation from the obligation to place its contracts for research through advertising for bids. This is particularly so since the measure of a successful research contract lies not in the dollar cost but in the qualitative and quantitative contribution which is made to our knowledge. The extent of this contribution in turn depends on the creative spirit and talent which can be brought to bear within a research laboratory. The National Research Foundation must, therefore, be free to place its research contracts or grants not only with those institutions which have a demonstrated research capacity but also with other institutions whose latent talent or creative atmosphere affords promise of research success.

As in the case of the research sponsored during the war by the Office of Scientific Research and Development, the research sponsored by the National Research Foundation should be conducted, in general, on an actual cost basis without profit to the institution receiving the research contract or grant.

There is one other matter which requires special mention. Since research does not fall within the category of normal commercial or procurement operations which are easily covered by the usual contractual relations, it is essential that certain statutory and regulatory fiscal requirements be waived in the case of research contractors. For example, the National Research Foundation should be authorized by legislation to make, modify, or amend contracts of all kinds with or without legal consideration, and without performance bonds. Similarly, advance payments should be allowed in the discretion of the Director of the Foundation when required. Finally, the normal vouchering requirements of the General Accounting Office with respect to detailed itemization or substantiation of vouchers submitted under cost contracts should be relaxed for research contractors. Adherence to the usual procedures in the case of research contracts will impair the efficiency of research operations and will needlessly increase the cost of the work to the Government. Without the broad authority along these lines which was contained in the First War Powers Act and its implementing Executive Orders, together with the special relaxation of vouchering requirements granted by the General Accounting Office, the Office of Scientific Research and Development would have been gravely handicapped in carrying on research on military matters during the war. Colleges and universities in which research will be conducted principally under contract with the Foundation are, unlike commercial institutions, not equipped to handle the detailed vouchering procedures and auditing technicalities which are required of the usual Government contractors.

Budget: Studies by the several committees provide a partial basis for making an estimate of the order of magnitude of the funds required to implement the proposed program. Clearly the program should grow in a healthy manner from modest beginnings. The following very rough estimates are given for the first year of operation after the Foundation is organized and operating, and for the fifth year of operation, when it is expected that the operations would have reached a fairly stable level:

Activity	Millions of dollars First year	Fifth year
Division of Medical Research	$5.0	$20.0
Division of Natural Sciences	10.0	50.0
Division of Scientific Personnel and Education	7.0	29.0
Division of National Defense	10.0	20.0
Division of Publications and Scientific Collaboration	.5	1.0
Administration	1.0	2.5
	33.5	122.5

ACTION BY CONGRESS

The National Research Foundation herein proposed meets the urgent need of the days ahead. The form of the organization suggested is the result of considerable deliberation. The form is important. The very successful pattern of organization of the National Advisory Committee for Aeronautics, which has promoted basic research on problems of flight during the past thirty years, has been carefully considered in proposing the method of appointment of Members of the Foundation and in defining their responsibilities. Moreover, whatever program is established it is vitally important that it satisfy the Five Fundamentals.

Legislation is necessary. It should be drafted with great care. Early action is imperative, however, if this nation is to meet the challenge of science and fully utilize the potentialities of science. On the wisdom with which we bring science to bear against the problems of the coming years depends in large measure our future as a nation.

9: RESEARCH ON MILITARY PROBLEMS

It is not necessary for me to expound at length on the importance of new weapons in modern warfare. That point is generally understood. Today it is evident to all thinking people that the evolution of new weapons may determine not only the outcome of battles, but even the total strategy of war. That has always been true to some extent, but today the rate of evolution of military weapons is much more rapid than it has ever been in the history of human conflict. Tomorrow the impact of new weapons may be even more decisive.

The great change in pace which science and technology have introduced into warfare underlines the vital importance of continuing an effective research on military problems in times of peace. In the past, the pace of war has been sufficiently slow so that this nation has never had to pay the full price of defeat for its lack of preparedness. Twice we have just got by because we were given time to prepare while others fought. In 1941 the margin was narrower than in 1914. The next time— and we must keep that eventuality in mind—we are not likely to be so fortunate.

The speed and surprise with which great damage could be done to our fleet at Pearl Harbor is only a mild warning of what might happen in the future. The new German flying bombs and rocket bombs, our own B-29, and the many electronic devices now in use which were unknown five years ago, are merely the forerunners of weapons which might possess overwhelming power, the ability to strike suddenly, without warning, and without any adequate means of protection or retaliation. I do not mean that some methods of protection or retaliation could not be developed. I mean only that we might not be given sufficient time within which to develop those means, once hostilities had begun, before disaster overtook us.

It is imperative, therefore, that we begin at once to prepare intelligently for the type of modern war which may confront us with great suddenness some time in the future. We all hope that no such event will occur. We all hope that means will be found to secure peace among nations, and we are anxious to do our full part in bringing about in due time an international organization and understanding that will truly preserve peace, but in the meantime we need to keep our powder dry. More specifically, we need to be effectively organized.

I think it is clear that we must not go back to either the organization or the philosophy which prevailed with regard to scientific research on military matters in the years between 1918 and 1939. Let me make myself fully clear. Many admirable things were done by both the Armed Services and civilian groups in those twenty-one years. Both the Army and the Navy, for example, can be proud of some real technical advances during that period. But that is not the whole story. To get a full picture we must remember also the neglect and the mistakes. Of these there was a full quota.

Yet, in my judgment, it is truly remarkable that the Services were able to accomplish as much as they did in the face of the tremendous obstacles which beset them on every hand in the peacetime years. The fundamental difficulty, of course, lay in the attitude of the American people toward preparedness for war. The American people were not prepared to build soundly for war during times of peace. We cannot, therefore, expect the Services to have accomplished what the people, by their attitude, made it impossible to do. I might add, at this point, that to the best of my own recollection, the Congress in those years regularly voted the research appropriations requested by the Services.

In addition to the attitude of the American people, however, there were at least two fundamental obstacles to a truly successful program for military research inherent within the Services themselves. First, and more important, was the internal organization of the Services. That organization gave insufficient recognition to science, its requirements or potentialities as a phase of warfare. Second, in addition to the organizational difficulties, service personnel, by training and tradition, did not, by and large, appreciate either the position which scientific research must occupy or the contribution it could make to any successful program for the national defense.

The essential fact is that we failed during peace to do as much as we most assuredly should. Certain of the reasons for that failure are obvious. They should be cured with all the vigor at our command. And it will require both vigor and courage, for the roots of the trouble are deep.

Since someone is certain to suggest that the answer lies in extending our wartime organization into the peace, let me meet that argument now. The argument has deceptive plausibility. It is deceptive for two reasons. In the first place, no temporary expedients, effective as they may be, can outlast the emergency pressures which gave them being and vitality. War improvisations should be recognized for what they are. In the second place, no temporary improvisation, excellent as it may be, can be completely effective if the fundamental organization upon which it is superimposed is either weak or unsound.

Under the pressures of war, our temporary expedients have worked well. It is not necessary for me to describe in detail these temporary expedients. The Office of Scientific Research and Development is one. It brought civilian scientists of the very highest calibre into active participation on matters of military research. It gave a civilian body, reporting directly to the President, authority and funds both to support and to *initiate* research on matters essential to the national defense. Within the Services there have been similar expedients. They have been explained by their responsible officers. The New Developments Division in the Army, under Brigadier General William A. Borden, and the Office of the Coordinator of Research and Development in the Navy, under Rear Admiral J. A.

Furer, are notable. Within the structure of the military itself, moreover, and reporting directly to the Joint Chiefs of Staff is the Joint Committee on New Weapons and Equipment, over which it has been my privilege to preside. This is the senior staff body within the military organization itself on new weapons and equipment.

When a new emergency arises it may well be necessary to create new emergency organizations for temporary action. There is a great advantage in new organizations created for explicit emergency purposes. They have not accumulated the rigidity and formalism toward which all organizations are likely to trend with the passing of time and when not under pressure. They are cut from whole cloth, and the vigor of youth applies to organizations as well as to individuals.

Yet the continuation of an emergency organization after the emergency has passed is likely to be as great an error as the failure to create it when needed. We should not mistake emergency remedies for a permanent cure.

Nor does the solution to the problem lie simply in the establishment of an agency through which the assistance of civilian non-Governmental scientists can be made available to the Army and Navy. The participation of civilian scientists in the scientific aspect of military problems is only one small portion of the total problem. The very heart of the problem of an adequate postwar organization for the conduct of research and development lies in the organization of the War and Navy Departments themselves. Unless this major problem is resolutely faced and affirmatively resolved, the solution of the peripheral problems (such as the form of organization for civilian participation in military research) will not enable us to prepare ourselves adequately for the wars which may come.

It may well be maintained that if the Army and the Navy are properly organized, and if they approach the problem of science and scientific research in the postwar era in a sound and thorough fashion, the enthusiastic cooperation of civilians will more readily follow. Even more important, a sound mili-

tary organization, which is receptive to the role of science and
of civilian scientists, will enable an auxiliary civilian organization to operate to maximum advantage. The converse is equally true. No mere addition of an auxiliary civilian body designed to aid the military effort will insure adequate development of weapons in time of peace if the military organization
itself is unsound or unreceptive.

This question of postwar organization for military research
and development is an integral part of the major question of
the over-all peacetime military organization of this country. It
seems to me that the surface of the problem of adequate or effective organization for the defense of this country has only
been scratched.

Since the over-all postwar military framework has not yet
been fixed, it is difficult for me to discuss the problem of organization for military research and development in any explicit way. Yet certain fundamental principles seem clear. In
the hope that it may be helpful, I should like to summarize
four of the important principles which, to my mind, should be
observed in any postwar organization for military research.

PLANNING AT THE TOP

(1) *There must be adequate planning at the top both for
the evolution of weapons and for the strategic use of new
weapons.*

At the outset, I should make it clear that in this discussion,
which will involve some criticism of the military system, I dif-
ferentiate between the system and the officers who operate
under it. There is no intention on my part to criticize any of
the officers in whom this country has placed its confidence and
who are among the most able military leaders this country has
ever produced. These officers and men have performed most
effectively under a system which is not calculated to make the
most of science and technology in modern war.

I also want it clear beyond all possibility of misunderstanding or misconstruction that, in my judgment, the military and

civilians, working together, fought the technical aspect of the late war in an effective partnership. On the scientific front, on research and new weapons, things in general went remarkably well when we consider the great organizational handicaps under which we started and which, to some extent, continued to persist.

Military tradition, for example, has in the past, called for planning in terms only of existing weapons. Such planning is done in view of all the complexities of logistics, training, intelligence, and personnel to fit into an over-all strategic program. The failure to have at the top levels of the military organization trained scientists and military leaders who plan in terms of future weapons or weapons in process of evolution may be costly in terms of lives and battles. Of course, planning for immediate campaigns must always be done in terms of existing weapons. The long range planning of a whole war, however, must go further. It must be done in the terms of the evolution of weapons and strategy. In the future, the presence or absence of this type of planning at a high level may determine the entire course of war.

Traditionally, the advanced military thinking on the improvement of weapons and on new methods of combat has been left to the lower echelons. The theory has been that any matter of sufficient importance will force itself up from below upon its own merits, and demand the attention it deserves. There is, of course, a basis for this argument. Yet, the course of modern war is so largely determined by the evolution of new techniques that it is absolutely essential that first-class thinking be done which combines military considerations with the possibilities opened by technical progress. This thinking can be done only at the top. To proceed without it in a modern war is simply not good enough. Progress in complex technical matters is slow enough in the face of inertia, limited vision, obstinacy, vested position, tradition, and all the other ills that flesh is heir to without superimposing the organizational handicap of making new thinking fight its way to the attention of the top level commanders who ought to be giving it affirmative consideration.

Lest it be thought that this is merely a matter of generalities, examples can best be drawn from the first World War. The three great technical innovations of that war were tanks, poison gas, and aircraft. All three produced effects on the course of the war, yet these were essentially temporary and local. If sufficient grasp and vision had been present to see possibilities at the outset and prepare for full-scale surprise use thoroughly followed up, there is little doubt that the war could have been shortened, one way or the other. In particular, tanks, in their then crude form, but in the absence of air opposition or anti-tank weapons, could undoubtedly have turned immobile trench warfare into a war of movement at that time as well as later, had they been exploited to the full instead of tentatively and on a shoestring. To have brought this about would have required the combined vision of military men with those who understood the numerous possibilities of tracked vehicles. The procedure whereby enthusiasts, with a novel method, convinced the High Command sufficiently to overcome skepticism and obtain a trial, could bring only small results; which it did.

There is need for technical planning at the top not only to give affirmative direction and drive to new developments and their use; there is an almost equal need for such planning at the top in order to coordinate the work of the several branches of the Services, both in the development and in the use of the new weapons in which more than one branch is directly interested.

Our present military and naval organizations were built for much more static armament than we have today. In the days of wooden ships and iron men it was not only sufficient, it was highly desirable, to place the full responsibility upon one officer to see to it that those ships were soundly built, and upon another to see that their guns were the best that could be constructed. We have come a long way from that situation, both in techniques and in organization; but our techniques have out-run our organization for handling them.

It is still true and it will always remain true, that for the effective performance of any organization, especially a military organization, responsibility must be definitely assigned and re-sponsibility and authority must go together. This sound prin-ciple should never be departed from. In the case of military research, however, this sound principle has been departed from by the Services themselves. Whenever a new research pro-gram or development appears which is not directly and entirely within the cognizance of a particular branch, there has been no body with responsibility and authority to direct and coordinate the work of the several branches. It is essential that there be such a person with power both affirmatively to direct new pro-grams and to integrate existing programs when they involve more than one Bureau or Service branch.

Modern weapons call for complex programs involving many skills—so much so that an item as humble as a flashlight if designed for special operational use may find itself the victim of overlapping jurisdictions and competing demands for control over both its development and procurement. Today, a single complex unit of offense or defense, such as a radar-controlled anti-aircraft installation, may cut across many of the traditional branches of the military service. A complex weapon on the or-der of the German V-bombs might, for example, involve Chemical Warfare if it were an incendiary, Air Forces if it were borne by wings, Signal Corps if it involved control devices, Engineers if it needed emplacements for launching, and Ord-nance. Similarly, a new type of incendiary bomb would in-volve both Chemical Warfare and the Air Forces. Each has a responsibility and each must be satisfied as to its own specifica-tions, standards of safety, and performance. No one branch sees the whole picture. No one branch can give over-all direction.

The requirements of modern war have outrun the erstwhile satisfactory formal organization. Bureaus and Services can have responsibility for parts of a complex development, instituted and ordered from the top, but often no one Bureau or Service can do the whole job itself. Nor in fact, can it be done from the top unless some of the men in positions of great authority grasp the trends of science and its implications. This they can do only if science and its applications have bulked large in their professional careers.

The problem, therefore, is to provide some means by which scientific and technical thinking of the highest calibre may fuse with military thinking at the top level of command. A number of things can and should be done to bring this about. The precise steps which must be taken will, of course, to a large extent depend upon the ultimate framework of the military organization. The full solution will, naturally, come slowly, particularly since the art of using scientific and technical thinking as a part of top level planning will not ensue merely by providing for its presence. There must be a conviction within the Services that individuals must be so placed. Moreover, such individuals must be of the intellectual fiber and background to enable them to synthesize the two types of thought, military and scientific, into an integrated whole.

TECHNICIANS IN UNIFORM

(2) *The position of the technical man in uniform must be improved.*

It is not enough merely that technical and scientific planning be done at the level of high command. The position of the technical man should be improved throughout the length and breadth of both the Services. Men in responsible positions should have better technical training. Conversely, soundly trained technical men should be eligible for high command. Lastly, broad or complex programs of research and development should have a status at a staff level.

Under conditions of modern war a grasp of broad technical trends would seem to be as fine a qualification, when combined with indoctrination in the art of command, as an officer of the top rank could possibly have. Yet, the practice of the two American Services seems to have been based on the assumption that technical specialization is incompatible with high command. It is hard to see how this tradition ever arose; yet it has certainly existed for a long period of time. It is reflected in many ways: in the limited courses in science given at West Point and Annapolis, in the failure of the regulations on promotion and selection for high command to give due credit to advanced scientific and technical accomplishment—in effect, the road to high command lies through field command alone. As an example, the Construction Corps of the Navy, although it certainly produced outstanding combat ships and although it attracted, while it existed, some of the finest minds in the Navy, nevertheless did not furnish a corresponding quota of officers to positions of high command. The difficulty under which the scientist labors is also reflected in the fact that research unsuccessfully conducted places a permanent blot on a military career. It should be recognized, however, that in research which is forward looking and advanced in nature, many failures are the price of one success.

Men in uniform should receive better and more nearly fundamental scientific training. Provision should be made for advanced scientific training of large numbers of officers either at special Service schools or through a fuller utilization of existing colleges and universities. The War College idea is a sound one. It should be extended to include an advanced military college devoted to the evolution of weapons and its relation to strategy. It should bring together for training officers of land, sea, and air. Its courses and problems should be tough so as to test and exert the best brain power. Successful completion of the course should give an officer definite credit toward promotion in the line that leads to great responsibility and high command.

The Service schools themselves will want to bring new methods of teaching and new subjects to their curricula. They will want to follow the modern trend and give more attention to expanding the horizons and broadening the mental grasp of the officers-to-be.

Until technical men in uniform are given better training and until they find a readier route to positions of command, it is certain that the top levels of our military command will not grasp the full implications of military innovations and will not be organized to handle them to optimum advantage in some possible future highly technical war. We have, of course, to-day some officers at high levels with a technical background and appreciation. The rapid technical advance of air warfare, and the constant peacetime association of the Navy with technical matters, inevitably has produced such individuals. My point, however, is that we must have many more such individuals and they must have a broader scientific preparation for their responsibilities.

The position of technical men in the Services suffers not only from the failure to give a position of importance and responsibility to skilled scientists and technicians, but also from the failure to give an independent and important status to research and technology itself. In the Services, research is subordinated as incidental to the work of branches whose primary interests and responsibilities are in other directions.

The Services have not yet learned—as industry was forced to learn a long time ago—that it is fatal to place a research organization under the production department. In the Services, it is still the procurement divisions who maintain the research organizations. The evils of this arrangement are many. Basically, research and procurement are incompatible. New developments are upsetting to procurement standards and procurement schedules. A procurement group is under the constant urge to regularize and standardize, particularly when funds are limited. Its primary function is to produce a sufficient supply of standard weapons for field use. Procurement units are judged, therefore, by production standards. Research, however, is the exploration of the unknown. It is speculative, uncertain. It cannot be standardized. It succeeds, moreover, in virtually direct proportion to its freedom from performance controls, production pressures, and traditional approaches.

In the case of research, a scientist agrees to use his best efforts in the solution of a particular problem and he is paid for the effort and talent which he devotes to the job. Since research is speculative, a research scientist must be paid—or promoted— whether or not he succeeds in solving the assigned problem. In the case of procurement, on the other hand, one must furnish a particular product to meet stated specifications and one is, therefore, paid—or promoted—only for a product which satisfies those specifications.

Moreover, a procurement unit is under constant pressure to produce sufficient quantities of material for use on the far-flung battlefronts and can make no adequate or special provision for the prompt supply in small quantities of important new devices. To be effective, new devices must be the responsibility of a group of enthusiasts whose attentions are undiluted by other and conflicting responsibilities. As it is now, with nearly the entire procurement scheme geared to the mass production of great quantities of standardized equipment, the small special job becomes an orphan.

The union in the Services of the research and procurement functions has one other unfortunate consequence. A procurement unit which also is responsible for research is not anxious either to use or to recognize the merit of developments made by those outside the unit. Human nature being what it is, and it is certainly no different inside military organizations from outside, the result is to slow down the adoption of devices which first appear or are first suggested outside of the procurement unit. This may be particularly serious when we remember that modern weapons may either draw their components from or be, at least in part, the responsibility of several competing procurement units—each of which is in a position to retard or advance the progress of the other.

What is required is a separate organization within the Services for research, for development, and for rapid procurement, in experimental production, of small lots of new equipment to be used for field testing, and in critical situations for actual use against the enemy. Such an organization must be in the hands of trained enthusiasts and, although linked at all levels with all branches of the Services, it should report directly to an officer on the very top level of command who has the training, vision, and competence to direct the broad formulation of new weapons and to devise the techniques by which they should be employed. This will make it a main staff function to coordinate research, procurement and requirements—a job which can be done only by men who thoroughly understand all three.

INTERLINKAGE BETWEEN THE SERVICES

(3) *There should be a genuine scientific interlinkage between the two Services.*

It is hard to realize that there was a time not so long ago when the two Services were completely insulated from each other in certain technical areas. The failure of the two Services to have technical cooperation at all levels was not only wasteful and short-sighted from the point of view of technical progress, but harmful to adequate preparation for the defense of the country.

This problem of technical interlinkage between the Services is, of course, only incidental to the important, broader problem of adequate interrelation between the Services on all subjects. That problem is one which should be explored very thoroughly. In whatever way adequate linkage between the Services may be brought about, whether by a permanent Chiefs of Staff organization or otherwise, it is evident that such interrelation must be extended to technical matters.

On this question of mechanics, however, it is my personal view that, in peacetime, the device of linking the two Services with joint boards which report only to the two Services will be ineffective. Such Boards would have no one in a position of authority to resolve the inevitable differences. The Boards could operate only by unanimous consent. In wartime, however, such an arrangement does manage to work. The reason, of course, lies in the fact that, under the exigencies of war, men will agree. In peace, there is no comparable stimulus to agreement. Moreover, operating on a basis of agreement through joint boards is contrary to the fundamental and ever valid military principle that some one person must have the responsibility, *and the power,* to resolve all differences.

The technical problems of the two Services are, of course, not the same. Yet, there are broad areas in which they overlap. The fact that the problems and points of view of the Services differ to some extent would be a distinct advantage in any interchange of ideas. This cross-fertilization of ideas between two groups, each with its own particular set of problems, has long been established as an essential prerequisite to successful research. Scientific achievement on one set of problems can often furnish the key to progress on a broad front.

Adequate interlinkage will also avoid unnecessary duplication of facilities and effort, particularly when it comes to building large numbers of expensive devices. This does not mean, however, that there should not be parallel programs within the two Services. In the early stages of research and development on any problem, parallel programs are essential, not only to insure that all avenues of attack are covered, but because a parallel approach affords the necessary stimulus of competition. It keeps research scientists on their toes. Parallel effort is often highly economical in a long-range sense.

Whatever interlinkage of the Services is ultimately provided, it should not be allowed to inhibit the *esprit de corps* and the pride in organization which is at the heart of much of our military strength. We owe too much to the aggressive fighting spirit of our Navy, of our Marines, of our Air Forces, and of our indomitable Infantry to take any steps which would reduce the justifiable pride of belonging to unique, aggressive, skilled, fighting organizations. The significance of the uniform should not be diluted in the name of economy. Rivalry between Services and branches, extending even as it sometimes does to absurdities, nevertheless is a real source of military strength.

(4) *There should be some form of partnership between civilian scientists and the military.*

An improved form of civilian collaboration with the military on matters of military research should be worked out. This may be done within the Services themselves or outside of the Services, by means of a civilian scientific body with both the power and the funds to initiate research. Both of these alternatives are worth a brief examination.

The problem of working out within the Services a status for civilian scientists and technical men is particularly acute. American practice has been to insert such civilians at various levels. As a result, civilians have had to report directly to the uniformed personnel at the level to which they were attached. In spite of notable individual exceptions, this practice does not attract into the Services scientific and technical men of the highest caliber.

American practice has also been to keep the civilian organizations attached to the Secretaries small. They operate primarily to keep the Secretaries informed and especially in connection with procurement, to implement the policies of the Secretary in connection with the business affairs of the department. The philosophy, and it is a sound one, is that the Secretary will determine internal policies and conduct external relations with other agencies of the Government, but will not interfere with the detailed performance of essentially military matters. Thus, in this country we do not find civilians operating important branches within the department and reporting to the Secretary.

British practice has been somewhat different. In the Admiralty, for example, there has long been a civilian structure, on matters of finance and business, reporting to the permanent Secretary of the Admiralty. Separate ministries for procurement divide the business from the military affairs of Army and Air Force up to the level of the Cabinet. In these circumstances, and under the pressure of war, there has been a decided trend toward the organization of research and developmental matters in such a way that civilian research workers report through civilian organizations to the top.

In this country during the late war the Office of Scientific Research and Development, an emergency agency, produced some of the same results. It was linked to the uniformed Services at all levels, but reported directly to the President. OSRD also maintained sufficient independence and initiative to insure vigorous action, and to hold its civilian workers under civilian control, which they understood and under which they worked best.

One solution to the problem thus lies in the great expansion of the offices of the Secretaries in time of peace, with a civilian branch devoted to research and development. This has its distinct advantages. It can be carried to the point where there is a full structure, on basic research, on far-reaching developments, in civilian hands, within the control of the department itself, but in parallel with the current improvement of existing weapons by the Bureaus and Services. The experience of the British, and their further development of this scheme after the war, will be worthy of serious study. The scheme is, however, contrary to what has long been prewar American practice.

Another approach lies in revised organization within the existing branches of the Services so as to achieve more effective use of their civilian employees. The emphasis should be on an essential professional partnership between scientists and military men. The conditions of civilian employment should approach, in opportunity for recognition, freedom of action, and group *esprit de corps*, those of the better university and industrial organizations, if scientific men of the highest caliber are to be attracted.

A third approach lies in expanding and improving the Services' system of contracting with private laboratories for research tasks. It is important that contractors be given a free hand to carry out their assigned tasks in accordance with their own conception of what should be done. In the last analysis, it is the effort and the particular talent of the skilled civilian scientists that should be sought. That effort and that talent will only suffer if the Services seek to substitute their own ideas and instructions or the ideas and instructions of their lower echelons for that of the skilled technical men in the contractor's employ. The research scientist is, after all, exploring the unknown. He cannot be subjected to strict controls, nor can he be given detailed instructions as to the solution of the problems which he is retained to solve.

Which of these or other approaches should be adopted by itself, or in combination with others, must depend in the last analysis on what our postwar military structure may turn out to be, on whether the position of the scientist within the Services is improved, and on whether the over-all organization for research and development within the Services is adequate.

Regardless of what the internal organization of the Services may be, it is desirable to have outside of the Services themselves a civilian body with authority, and funds, to conduct research on matters which have military significance.

Such a body would supplement the work of the Services with the freshness of approach and independence of mind which are invaluable to successful research. For nearly thirty years the National Advisory Committee for Aeronautics has been such a body. That Committee, of which it was my privilege to be Chairman for several years, supervises and directs the scientific study of the problems of flight in the closest of cooperation with both the Army and the Navy. If it had not been for this Committee, our country would not have been in the strong position in which it found itself in aeronautical matters at the beginning of the late war.

We must face up to the fact that although research on military problems and military weapons is largely a military matter, it is not entirely so. The job is primarily one for the trained professional scientist and engineer. It is a mistake to believe that since science has military importance, scientific research should be run exclusively by military men.

The two Services exist to fight. That is their primary reason for being. In order to fight a modern war, the military must draw upon industry, agriculture, science, and all the other facets of our economy. Because industry, agriculture, and science have vital military use does not mean, however, that they should be made the exclusive responsibility of the military.

It is true that the Services have been charged with the defense of our country. It does not follow from that premise, however, that all aspects of the civilian economy which have vital military significance should be the exclusive function of the military. Not only would that be illogical, it would be a great practical mistake. The military cannot be expected to be experts in all the complicated fields which make it possible for a great nation to fight successfully a total war. The job of fundamental research can succeed best if given to those best trained to handle it.

Civilian science must clearly do the job, which by specialized training it is equipped to do. Civilian science cannot make its true contribution, however, if its efforts are subject to the complete direction of the military or if it has no independent funds. The real answer to the problem, of course, is a partnership between the military and civilian scientists. But a *true* and *effective* partnership can come about only if both are equals in a common endeavor. They must be equals and independent in authority, in prestige, and in funds.

I am so sure that the evolution of weapons is exceedingly important in the conduct of modern war, so anxious to impress this importance that we may be fully prepared for what may lie ahead of us in the future, that I should add a word to be sure I am not misunderstood. The weapon is an adjunct and a tool only. It should be better than the tool in the hands of the enemy. But wars are fought by men. The fighting strength of this country rests on many factors, and we will be strong only as each element is strong. It rests on the skill of military leaders, their ability to command, their readiness to accept appalling responsibility which would crush men of weaker fiber, their knowledge of their complex profession. It rests on business men, on their patriotic willingness to take chances in a common cause, on their ability to organize and manage complex affairs. It rests upon every laborer at the bench or in the field, upon their technical skills and their determination to support their sons in combat by supplying them fully. It rests upon the women of the country, as they endure the stress of war, and as they encourage and support their men folk, and carry on men's work. It rests especially upon the men in the ranks, upon their ruggedness of body and mind, upon their fighting spirit, upon their belief in their cause as they come to define it.

The scientists in war are merely one more group in a country which fights. They have special skills and training, which can be especially useful when intelligently applied, and which should not be squandered. They are in no sense a privileged class; they are just one more group in a democratic society, ready and willing to put forth their best efforts in whatever way will best serve the common cause. They are exceedingly proud to have been full partners in the dangerous effort just past, and they will stand ready to serve their country in peace by adding their bit to the national effort to maintain this country strong.

10: THE CONTROL OF ATOMIC ENERGY

Development of systems for the control and utilization of atomic energy is the most important task ever faced by the governments of the world. For the continued progress of civilization, it is imperative that people be safeguarded against sudden destruction by atomic bombs. It is highly to be desired, for the betterment of living for mankind everywhere, that the great resources of useful power offered by further development of atomic science become generally available.

No more intricate and exacting problem was ever posed to governments than this one. It is inherently complex because the science of the atom is complex. The fact that the deadly military potentialities of the atomic bomb and the beneficent industrial applications of atomic power are almost inextricably intermixed complicates it further. The urge to prevent wars is very strong in all minds, for we have just emerged from a terrible war. The desire to enjoy the better life promised by applications of atomic power is strong also. Because the means of producing this peaceful power can readily be converted into an atomic bomb for destruction, the mechanism for world peace and the mechanism for world control of atomic energy are profoundly interrelated. Preventing war is a long task, which must be done bit by bit; so also is the development of peaceful atomic power. The two must be related in our thinking.

In a nationalistic world, all peoples will seek to attain an equal footing with respect to anything so powerful as atomic energy. If the mechanism of world peace is available and is strong enough, peoples may be expected to relinquish something of their traditional nationalism to attain that equal footing through international organization. If no dependable mechanism is available, a secret arms race in the surreptitious development of atomic bombs may be foreseen. Any well thought out plan for orderly progress toward the banishment of war will start here, with the prevention of such secret preparations. The end of open preparations may then follow, to be followed in turn by the end to war itself. The start of the journey toward this great goal is in such a seemingly simple thing as the establishment of the complete flow of information—particularly of basic scientific information—across national boundaries.

The three governments which shared knowledge and skill to enable scientists to achieve the chain reaction and industry to create the materials for the bombs that ended the war have fittingly made the first move toward the establishment of mechanisms for the control of atomic energy and for the development of peace for which that control is an essential requirement. The declaration which grew out of the recent conference of President Truman, Prime Minister Attlee, and Prime Minister King is a very important document. It would have been important simply as a declaration for a peaceful world. It is of the very greatest importance because it chose the right path to the goal, blocked out the journey into practical, sensible stages, and clearly mapped the crucial first marches. The declaration is notable moreover because it entrusts to the United Nations Organization this momentous international responsibility. It is of the utmost importance that the member states, especially the great powers, upon whose co-operative effort success depends, do all within their capacity to assure that success. The way to international collaboration and control has been opened. But it will be a long way. As we progress along it, the separate national states must set their own houses in order, establish internal systems of control, and thus bring into being agencies which can support the international agency.

Eagerness to assure world peace is laudable enough. The

great hazard is the kind of over-eagerness which cannot endure the long patient work which will be needed, and which therefore argues for quick answers, such as "outlawing" the atomic bomb. Premature outlawry could well be disastrous, for it is impossible to outlaw when there is no effectively supported law. The first task is to create this. A good start has been made. The best possible support which our government can bring to that good start is the expeditious passage of sound legislation for domestic regulation and development of atomic energy. During the lengthy period necessary for creation of an international system, it will be possible for this government not only to enact the needed legislation but also to operate, test, and if necessary revise the domestic control system which that legislation establishes. By the experience of so doing, we may well secure experimental evidence that will be helpful in the performance of the international task.

Moreover, the passage of suitable domestic control legislation is urgently needed for purely domestic reasons. The present state of world affairs demands a strong United States. This is no time to let delay dissipate our strength, to let doubt and indecision hamper the great program of atomic science on which we are well embarked. People are getting tired of hearing about the atom, and when people get tired, they tend to turn away from issues. This is an issue which cannot in conscience be so ignored. Between the first and second World Wars, the United States experimented with disarmament in an unorganized world—to sad result. The sort of thinking which that involved should long ago have been discredited. If we are to do our fair share in the exacting, patient work of international regulation, we must vigorously get at the domestic task now. Though the problem itself is great and new, it is susceptible of solution by the same means which have brought governments into being, gradually rendered them more efficient as agencies for the general betterment of man's lot, and reached a high point in the United Nations Charter. Those means are hardheaded analysis and honest good will. The first, used to the full in determining our internal control system, will clear the way for the second in international affairs.

From this point of view, then, let us consider general principles which legislation for internal control and development of atomic energy should embody.

The vast physical plants, the stockpiles of materials, the varied applications of knowledge which taken together constitute the Manhattan Engineer District belong not to any man or group of men, not to any corporation or group of corporations. They are the property of the people of the United States through their government, in which are vested title to the physical properties and patent rights covering the engineering processes. This is as it should be, for the power of this development, for good or for ill, is too great to be otherwise held. Legislation for the further control and the further extension of this development should fortify this condition. At the same time, it should make proper provision for the active participation of private individuals and private corporations in the further utilization of atomic energy, under sound regulatory procedures embodied in a sensible licensing system.

For the years immediately before us, the deadly rather than the beneficent power of atomic energy will continue to hold first place in men's minds. Until suitable and effectual means for international control of military applications of atomic energy have been established and proved, the atomic bomb will continue to be a menace. Rigorous provisions for security concerning military exploitation of atomic energy, therefore, will continue to be essential. Our experience here will be of double value in the international effort, for it can become the basis for provisions to eliminate from the war machinery of nations not only atomic weapons but also other weapons nearly as deadly, by which we should be seriously imperiled in another war if atomic bombs had never existed. To solve the problem of the bomb is important in itself, and of greater importance still as a contribution toward solving the entire international problem of war.

The manufacture of fissionable materials is by long odds the most dangerous manufacturing process in which men have ever engaged. The process is accompanied by the production of radioactive by-products as poisonous as the basic material itself; should the process used in producing power be ill managed and get out of hand it would produce a great and deadly volume of such poisons. Improper or incautious manipulating of substantial amounts of fissionable materials by inadequately trained or irresponsible investigators is a danger to the public safety which government must avert. Legislation for the internal control of atomic energy would be shortsighted indeed if it did not make thorough provision in this regard.

No better illustration of the complexity of the atomic energy problem can be had than the dilemma posed by this need for security and public-safety provisions as against the need for scientific freedom in the further study and investigation of atomic science. Atomic energy as we now know it and as we have employed it in the making of atomic bombs came about as a result of long and patient experimental investigation. If we are to press forward with the further development of atomic science for employment in peaceful power installations, if we are to explore to the full the beneficial possibilities in medical use of the radioactive by-products of the fission process, if we are to go ahead with the search for pure knowledge in the field of which our present atomic knowledge is but a small part, we must so arrange controls that the research worker will possess the right and the freedom to carry on his studies.

A nice line of distinction hence must be drawn in security and safety provisions, in order to make possible not only untrammeled investigation but also the publication and sharing of results through which alone can we be sure of the fully productive thought on which scientific advances are based. The best way of determining this line is to define the critical situation that trenches on safety or security and provide for regulation up to that point only.

The United States is on record in favor of open doors in laboratories throughout the world and has declared its readiness to open its own doors if others will do likewise. Our legislation must be so drawn as to give substance to this declaration. Study of safety and security provisions must in addition take into consideration such questions as whether we should go beyond the declaration and open our laboratories before we are sure this is world policy, and whether we should provide for dissemination of our findings regardless of reciprocation.

Extreme care in the formulation of the legislation, extreme judiciousness in the selection of men to administer the legislation once drawn, are essential if we are to insure against freezing the science at its present stage, hamstringing further study, and repelling able minds from this field.

Not at once, but surely in a reasonable time, it will be practicable for man to use the controlled energy of the atom for direct peaceful purposes. Atomic energy as a source of steam power and of electric power will in due course—and not necessarily a long course—become available. If only because the best way to insure against the existence of atomic bombs is to separate and distribute their components in industrial installations, we should seek to bring about the industrial use of atomic power at the earliest moment.

The foregoing five principles, it seems to me, must be properly recognized in legislation for the domestic control and development of atomic science. Such legislation must insure to the American people their control of plant and process, must safeguard knowledge of the military applications of atomic energy, must properly guard the physical well-being of the people against the many hazards to life and health which the investigation and production of atomic energy involve, must provide for free and full research and interchange of knowledge in this new and promising field, and must reckon with the future task of putting fissionable materials to useful work.

11: RESEARCH AND THE WAR EFFORT

One of the advantages of a democracy is that when it is engaged in a war, no one feels that everything should be controlled by the military. There are great areas where civilian organizations can operate to better advantage, and this is and has been our accepted policy. The joints may creak at times, and there is bound to be confusion simply in view of the enormous magnitude of the job involved, but in general we get along faster when civilian organization produces the weapons with which the Army and Navy fight. The same advantages may be cited for civilian research and development, collaborating closely with the armed services, and meeting their needs as far as is physically possible, but acting with that flexibility and freedom which come from independent organization.

I believe, and I know that I am joined in this belief by most of the men with whom I have worked closely in the past few years, that we got on more rapidly and more effectively during the late war in the development and introduction of new weapons under the form of organization whereby civilian groups supplemented the work of the Army and Navy than we would have had the entire affair been closely under military control. There are many reasons for this. One of them is the fact that the Army and Navy were exceedingly busy with immediate considerations. It would be difficult indeed for a military organization to provide adequately for the long-range view while at the same time carrying its enormous responsibilities in regard to the battle which may come in a few months.

I feel sure that new and valuable ideas are much more likely to come to fruition if they can develop their formative stages among groups of independent scientists and engineers before being subjected to the rigors of military association. When we are engaged with skillful and resourceful enemies, we should not at any point underestimate them.

Certainly Germany had too long a history of scientific and technical accomplishment for us to underrate its possibilities in applying its skill to the conduct of war. It is no secret that Germany was fully engaged in the development of war techniques for a much longer time than the democracies. Nevertheless, I believe that the rigid military regime in Germany was at a disadvantage, when it came to the development of really new ideas, as compared with the United States, with its ingenuity and resources. I believe that this is especially true in view of the fact that the organization under which we operated gave full rein to the independent efforts of some of the finest scientists and engineers that the country has produced, under conditions in which they worked substantially in their own way and in accordance with their chosen methods, but toward a common end.

But a democracy in wartime has certain handicaps also. Unfortunately, from one point of view, our own country has a striking difficulty in adapting itself to modern war which is not generally realized. We are a people who think in terms of mass production. This is excellent, and it constitutes one of our greatest factors of strength. It has, however, the distinct liability that we are likely to think in terms of freezing of designs and production of great masses of standardized equipment, and we think much less readily in terms of a rapidly changing technical situation. In modern war it is a serious thing to be inflexible in this regard.

Weapons are continually changing. Some years ago I used to meet military men who took the point of view that once war was entered upon, the fighting would have to be done by utilizing the instrumentalities available at the beginning of the conflict. I have not heard this view expressed for quite a long while now. Every phase of warfare is changing and is changing radically, and the change is coming about primarily because the methods used are vastly different.   The central problem, therefore, in the effective conduct of war from this standpoint is to be sure that our weapons are thoroughly up to date.   This involves a long chain of endeavor, beginning with scientific research and engineering development, and proceeding through tests, procurement, installation, and the training of personnel, to the final use in combat.   If any one of these steps is not thoroughly and carefully taken, the end result will not be sufficient.

But the process must be swift also, and mass mindedness is a dangerous state for us unless we also keenly realize its dangers.

Under ordinary peacetime circumstances the progress from a brand new idea to its use in quantity by the public occupies at least five years.   There has to be research, development, and engineering design.   There has to be design for production and user experience obtained under carefully controlled conditions. Out of this can come a well-engineered device adapted for production in quantity to meet a mass need in an economic manner.   Under ordinary peacetime conditions a company that is introducing a new product will short-circuit this proper and deliberate method at its peril, for large indeed are the penalties of plunging into quantity production before all of the loose ends are tucked in.   Yet in time of war we are faced with the dilemma of shortening this process or else being dangerously behind the times.   Under the stress of war it is possible to compress the time scale somewhat.   If it is compressed too much and proper engineering is not accomplished, the results may be very sad.   On the other hand, delay in getting new devices into operation in time may have consequences that are disastrous.   The attainment of a proper balance in this regard is one of the most difficult problems confronting the industry that produces the devices and the military groups that utilize them. I feel that on the whole we have done a remarkable job of attaining a just balance and I wish to emphasize strongly that the matter can never receive too much consideration and attention in a country such as ours, where all our normal peacetime habits lie along the lines of standardization and large-scale production.

In the progress of a new weapon, from the first idea to the final use, engineers and scientific men of professional grade enter at many points. Notably they appear as a part of the personnel of the armed services themselves, and they appear also in' those services which are auxiliary to manufacturing effort but none the less essential if the whole scheme is to function adequately. I shall not attempt to trace all aspects of this matter by any means. I feel, however, that it will be worth while to trace the phase that is concerned primarily with the development of the new weapon from the standpoint of the government organization which was charged with this responsibility in the late war.

The scientists and engineers of this country were organized under governmental auspices for the development of new weapons when in June, 1940, there was formed the National Defense Research Committee, charged by the President with the duty of research and development of new weapons and instruments of war. The initial organization was relatively small, but it grew to a considerable scale. In June, 1941, a reorganization occurred and the Office of Scientific Research and Development was formed by executive order. OSRD was given the broad task of coordinating the efforts of scientists and technical men in connection with many phases of the war effort, but it was also given the definite charge of pursuing aggressively the work that had already been started by NDRC; and for this purpose NDRC was incorporated into its organization. At the same time OSRD also was charged with the carrying on of medical research closely associated with the prosecution of the war, and it did so through the efforts of the Committee on Medical Research which was a part of its organization.

Toward the end of the war, OSRD was making expenditures at the rate of about $175,000,000 a year. In terms of the over-all war cost this is not a large sum of money. It is, however, a substantial sum when considered in terms of research and development. OSRD operated entirely by contracts with existing academic institutions, industrial organizations, and government agencies. This method was designed to utilize to the utmost available facilities and personnel and to avoid, as far as it could be accomplished, the construction of great new laboratories. About 2,260 contracts for the carrying on of research were entered into. Approximately 315 industrial laboratories and 150 colleges and universities were engaged in work on OSRD projects, and the number of men of professional grade involved was in the neighborhood of 6,000.

The way in which the organization functioned will, I think, be of interest. There was the closest sort of interrelationship with the Army and Navy at all levels. NDRC was broken down into 18 major divisions concerned with various phases of war research, and in many cases these in turn were subdivided into sections. Each section was composed of scientists and engineers who were specialists on some phase of the enormous range of war instruments. Working closely with them were officers from the Army and Navy who were also specialists in the fields concerned, but who had in addition the war experience and the contact with tactical reasoning which is essential for sound planning.

The research projects arose in the sections, usually by reason of round-the-table discussion of the changing situation and the need for improvement. Out of such discussions came usually a definite request from either the Army or the Navy that OSRD undertake a development along a certain line. The section was charged with the duty of finding the best laboratory for the conduct of the work and the best personnel to carry it on. After these considerations were met, the section then recommended a contract for the accomplishment of the work. This recommendation was reviewed by NDRC, and if the Committee approved, it passed the recommendation, together with its endorsement, to the director of OSRD, who authorized the work to proceed.

The NDRC examined the operations of divisions through the medium of small subcommittees, and by bringing before it the chiefs of the several divisions.

The manner in which NDRC operated in this connection was of great importance since it was the central reviewing agency which tied together the entire program. Dr. James Conant, president of Harvard University, was the chairman of NDRC, and its members were as follows: Dr. Roger Adams, head of the department of chemistry of the University of Illinois; Dr. Karl T. Compton, president of the Massachusetts Institute of Technology; Dr. Frank B. Jewett, president of the National Academy of Sciences; Dr. Richard C. Tolman, dean of the graduate school of the California Institute of Technology.

These men primarily represented American science and engineering. Conway P. Coe, then Commissioner of Patents, was also a member; he brought to NDRC wide knowledge of inventions and their appropriate handling. Notable among the Service representatives were Major General C. C. Williams, whose office in the War Department's Services of Supply was in touch with developmental work throughout the Army and was also in touch with the needs of the Army; and Captain Lybrand P. Smith of the Navy Department, where he served in the Office of the Co-ordinator of Research and Development.

Once a project had been authorized by OSRD in the manner described above, the office of the chairman of NDRC was charged with the duty of administering the project from its scientific and technical standpoint. For this purpose the line of authority flowed from the director of OSRD, through the chairman of NDRC, to the divisions, and members of these divisions became the authorized representatives of the office in the guidance of the contractors in their scientific and technical research, in order that their efforts could be directed along the lines that had been approved by NDRC as best adapted to the needs of the services. In this way also adequate reports of progress were made available promptly to the interested parties in the Services.

Since OSRD was concerned with many broad aspects of the relationship between the military Services and the civilian organizations, the director of OSRD also had the benefit of an advisory council representative of many points of view. The council comprised: Harvey H. Bundy, special assistant to the Secretary of War; Rear Admiral J. A. Furer, co-ordinator of research and development of the Navy; Dr. James B. Conant, representing NDRC; Dr. A. Newton Richards, chairman of the Committee on Medical Research; and Dr. J. C. Hunsaker, chairman of the National Advisory Committee for Aeronautics.

It also included in its discussions Dr. Harvey N. Davis, director of the Office of Production Research and Development of the War Production Board. By special direction of the President, the director of OSRD also had the benefit of the advice of the president of the National Academy of Sciences, who joined in many council deliberations. The work of all of these organizations will be discussed later, for their interrelation in the technical phases of the war effort is of much importance.

To continue, however, with the actual functioning of OSRD, I wish to mention several other phases of its activities and some of the problems that it faced. The business affairs, concerned with contracts and the like, were handled by the executive secretary of OSRD, Dr. Irvin Stewart, who also conducted relations with other governmental agencies on financial and legal matters.

It was a task of no small magnitude so to fit OSRD into the framework of government that it could operate in a smooth and effective fashion. Throughout its existence, the agency was a part of the Office for Emergency Management, which was in the executive office of the President. OSRD attempted, and I believe with extraordinary success, to carry on its affairs strictly within the framework as laid down by Congress, and in accordance with the regulations for the conduct of government business with which these various agencies were charged.

Throughout the rapid growth of OSRD it had exceedingly effective support from all of the agencies with which it necessarily came into contact, notably with the Bureau of the Budget, the General Accounting Office, the Civil Service Commission, and many other groups with which it was concerned as an independent agency within the executive office of the President. It is a pleasure to report that throughout my experience as the head of a new and vigorous government agency, I never met with anything but the most helpful attitude on the part of the agencies with which I was called upon to deal, and with the committees of Congress that had to do with the affairs of OSRD. This was in no small degree due to the excellent support of the executive secretary of the agency.

OSRD included also a liaison office, responsible to the director and charged primarily with the duty of conducting appropriate technical interchange with the allies of the United States. Under specific instructions from the President, there was instituted very early a close interchange with the British on technical matters and this relationship continued in a cordial and effective manner. I feel sure that this interchange expedited the work of scientists and technical men in England in their magnificent efforts for the protection of the British Isles, and I am sure that it benefited the United States in its war effort. The liaison office served the London mission of OSRD through which there were contacts with the British Government at all times.

The close interrelationship of science and engineering is essential in the early aspects of the development of a new weapon. Accordingly, the divisions of NDRC were made up of men chosen from both fields, working closely in collaboration. Incidentally, these men were selected both from universities and from industry, from large colleges and small colleges and from private laboratories, and they were drawn from all over the country. Many of them served without remuneration on a part-time basis; others were on the government payroll while on leave of absence from their organizations. They served in every case, of course, as individuals, and they were chosen for their individual qualifications.

In the introduction of a war weapon into use there is a special problem which is unique and which is not encountered in the same form by industry in the course of its development of new devices. During the course of the introduction of a new weapon it passed from the hands of OSRD directly into the hands of the armed services. OSRD was charged with the research and development, but it was not charged with procurement and use, which were in the hands of the armed services themselves. Needless to say, the armed services themselves directly and through contract carried on a great deal of research and development, and many of their new devices came through their own channels. This I will mention again later. However, at this point I wish to trace the handling of the problem which occurred by reason of the transition of devices from the laboratory into the hands of the military. In order to coordinate this aspect of its work, NDRC maintained two special pieces of organization. One was an engineering panel, made up of engineers who were at the same time members of the various divisions, together with certain other engineers chosen for their over-all grasp. This panel was charged with the duty of seeing to it that appropriate engineering skill was made available to the divisions and sections in an effective way at such time as a new device approached the period in its development where it began to be adopted for actual production and use.

The other special piece of organization was the so-called transition office, which was charged with the responsibility of following the progress of devices, in order to make certain that the problem of scarce and strategic materials was considered in sufficient time. The transition office also arranged with the armed services for initial production in order to carry the device through the transitional phase, in which it emerged from the laboratory but did not appear in quantity. At this point there was usually involved the production of a sufficient number of pieces of equipment, often produced by hand methods, for purposes of extended tests in the field. There were involved also the selection and indoctrination of an appropriate manufacturer.

CONFIDENTIAL WORK OF OSRD

In order to appreciate the way in which OSRD operated, it is necessary to realize that practically everything it did was highly secret. Inevitably, it was not possible to carry on work under conditions of great secrecy with the same dispatch which is possible when no such conditions obtain. For reasons of security, appointment of personnel in any capacity throughout the organization was made only after careful investigation. A ruling principle, and one which was observed by the Army and the Navy, was that secret matters were to be held carefully in compartments. This meant that no member of the organization could learn of secret matters except to the extent that was necessary for his appropriate functioning in the particular position which he occupied in the organization. Knowledge concerning especially secret matters was restricted to decidedly small groups within OSRD and within the Services themselves.

This leads me to mention one other matter. Throughout OSRD's existence, I encountered many times the question as to why it had to be organized on a national and vertical basis in accordance with subject matter, and why it could not be decentralized geographically to obtain the benefit of the many individuals in the country who were highly capable in technical ways, but who necessarily had to operate in their own localities. The necessity for secrecy and compartmentalization was the reason. In many cities, it would have been quite possible to form very strong technical and scientific groups locally, composed of men able to put in part of their time, in the evenings and on week ends, on technical matters connected with the war. These groups could represent many sciences and many types of engineering, and they would have been made up of decidedly effective individuals. However, this scheme was not compatible with tight restrictions on the work of OSRD. It would hardly have been possible to assign one subject to each group in a locality. Neither would it have been possible to give to such a group the knowledge of the entire range of the development of weapons which would be essential in order to use effectively the diverse characteristics such a group would have. Hence we reluctantly felt that such groups could not be utilized in the affairs of OSRD. However, they could have real value in other connections not involving stringent conditions of secrecy.

On the other hand, while OSRD was organized nationally, drawing its membership from all over the country, its sections were made up of men especially adapted for the problems before them and these men were given full knowledge of the technical and tactical phases of the particular weapons with which they dealt. They were kept closely in touch with the progress made in introducing certain weapons in practice, and they formed teams which were able to enlist the services of large numbers of men in many universities and industries for the accomplishment of their purposes. All this was done in such a manner as to keep secret information as closely confined as was consistent with rapid progress.

As previously ment   ed, not all of the research and development on weapons in this country was carried on by OSRD. It was the duty of OSRD to relieve the Armed Services as far as possible in this regard, and indeed as the war proceeded and as the officers in the Services became more and more burdened with immediate matters concerning the conduct of the war, the load in regard to research and development shifted so that OSRD gradually carried a greater share of the burden. However, it should be noted that both Armed Services maintain large laboratories in peace and in war for the development of weapons, and they also further development by direct contract with industry.

The mention of this point gives me an opportunity to discuss a matter about which there is—and was—a great deal of misunderstanding. It has been publicly known for a long time now that when Germany started its all-out air attack on Britain in the summer of 1940, the attack was repelled not only on account of the magnificent equipment and fighting qualities of the Royal Air Force, but also because the British had and effectively used radar as a means of taking the surprise out of the Germans' attacks and assuring that their bombers were promptly met by fighter squadrons. It is also known that the British had this device because of the effective work of a group of British scientists and engineers over a considerable period of time. I am also very glad to be able to state that at the same time the Army and Navy of the United States had equally effective devices for this purpose, well developed and in hand. This had been accomplished during years of peace, in spite of the fact that the United States had failed to support its military departments to an extent which rendered research and development in peacetime possible on anywhere near an adequate scale. In particular, I know personally of the early work in this field by a small group of keen naval officers, and there were undoubtedly other groups at work elsewhere. I also wish to emphasize strongly that this work was done long before Europe went to war, still longer before there was any such thing as NDRC. Work along the same lines was, of course, pursued by NDRC in the course of its existence. In so doing, it was proud to collaborate with the Army and Navy and to work in partnership for the further development of devices on which they had already pioneered, and to share in all of the various possibilities flowing out of that early work.

MILITARY RECEPTIVITY TO CHANGE

This leads me to a matter that I have pondered for some time. I am occasionally met by the old accusation that military men are hidebound and reactionary and that they are generally resistant to the introduction of new ideas. As applied to the wartime Army of the United States, such a statement is of course absurd on its face, since 24 out of 25 officers came directly from civilian life. The community of our officers during the war was essentially a cross section of the general public. However, the statement has usually been made with respect to the Army and Navy men who have made military matters their profession.

In the course of my duties as the head of OSRD, I worked very closely indeed with a large number of Army and Navy officers. Prior to that time I had had long association with engineers in the United States, with college faculties, and with businessmen. In every one of these groups, I have met individuals whose receptivity to new ideas was absolutely zero. I shall not single out any particular group for comment, but in every one of the great sections of our population I have been struck at times by an unwillingness to recognize the changing nature of the technical world to an extent that annoyed me exceedingly. I have seen it among college teachers and in groups of engineers, and I have certainly seen it among businessmen. I have also seen the same thing among military officers.

However, I can say categorically—and I think I am in a position to know—that the men who lead the Army and Navy of the United States are no less open to new ideas than is the general public of the United States. And if we as a country are over conservative and disinclined to try new things, then I do not know what the words mean. True, officers who have seen long service are hard-boiled. They have a keen appreciation of what will and will not work—at sea under difficult conditions and various types of weather, on land in the dust and mud of battle. They are intensely practical and exceedingly busy men, but they are not reactionary. In five years of close association with them in the development of new weapons, I did not see an instance in which an idea or a device which had come to the attention of OSRD and which in my opinion had outstanding merit was turned down arbitrarily and blocked permanently by any military officer or any group of officers.

The merit of new ideas has to be proved. Ideas have to go through their growing periods and meet their stresses. This has to occur in business in time of peace and in military affairs in time of war, but the atmosphere is no more hostile in one case than it is in the other.

OTHER DEVELOPMENT AGENCIES

But to return to civilian organizations, there were several aspects of the development of new weapons which very definitely did not come under the control of OSRD, although our council functioned in an advisory capacity in order to provide a unitary approach to problems of common interest and to prevent overlapping and duplication.

NATIONAL ADVISORY COMMITTEE FOR AERONAUTICS

Notable in this connection was the National Advisory Committee for Aeronautics. Founded by Congress in 1915, it has a long and significant record of accomplishment. Since the problems of flight were being attacked adequately by the NACA, they were not again attacked by NDRC, although the latter often carried on work on military devices which became incorporated in airplanes. It is a fortunate thing for the United States that it has had for many years the benefit of the active research of the NACA in the field of aeronautics, carried on in close collaboration with the Army and Navy and with industry. Early in the war there was a great deal of discussion as to whether American airplanes were comparable with those of the enemy. When the records began coming in from England, Africa, and the South Pacific, this discussion was quickly resolved. Because of the interaction of many factors, but particularly because of the fact that we have long had the NACA as an active independent research organization working on a basis of excellent interchange with the Army and Navy and supplying the fundamental basis for the advanced design of aircraft, this country has not lagged behind in air power.

NATIONAL INVENTORS COUNCIL

The wartime ideas that finally became incorporated in new military devices originated in a great variety of ways. Many of them came directly from officers of the armed services—this was increasingly true as our combat men called attention to their needs and opportunities. Some ideas came from the scientific and military groups assembled as sections of NDRC, as a result of conferences and discussions. A large number were submitted by the general public. However, the percentage of valuable suggestions coming in spontaneously from the lay public was necessarily small, all things considered, and required a great deal of review. This review was provided by the National Inventors Council, located in the Department of Commerce. There are one or two points in regard to the work of the NIC which I should like to discuss.

It should be emphasized that the NIC was the official reviewing agency and that its function was fully performed when it brought a valuable suggestion to the appropriate attention in the armed services. At the same time it is essential to emphasize that OSRD did not have the duty of reviewing suggestions submitted by the general public. When the armed services found that an idea warranted development, they sometimes turned to OSRD in order to have such development performed. But while research and development were the main functions of OSRD—and it attempted to stay strictly within those bounds—those sections that were making plans for the development of new weapons received many ideas from their immediate personnel, from engineers and scientific men working with contractors who were carrying out research under the supervision of the section, from the officers and men of the armed services, and through the armed services from the Inventors Council.

Unfortunately, the independent inventor was at a very considerable disadvantage when it came to the matter of making valuable suggestions in connection with military devices. In the nature of things, he could not be told the entire state of the art to which he was attempting to contribute, so that he worked very largely in the dark. This was unfortunate, for the same reasoning was unduly gone through over and over, but it was inevitable in view of the necessity for security. When a man who had put considerable effort on the development of an idea finally submitted it, he naturally wanted to know whether his idea was new, whether it was considered valuable, and, especially, if it was being used. Yet in general he could not be told. Indeed, he remained completely in the dark unless he occupied a position in which he was entitled to secret and confidential information in the field of his inquiry. If this principle had not been strictly adhered to, the enemy might have found out a great deal by simply putting in suggestions and thus learning the general state of the art and the status of development of various military weapons.

I mention this because I believe there has been a great deal of misunderstanding on the matter. Many individuals were distinctly annoyed because they were not told of the outcome of review when they made suggestions. It is strange that this situation was not more fully appreciated. To cite an interesting example, a prominent magazine ran an article describing in detail a military device which had been submitted to the armed services. The article complained bitterly that no serious attention was paid to the suggestion. As a matter of fact, the inventor in this particular instance had visited me personally and described his device. At the time he described it to me, I knew that a better device than the one he suggested was already in use. This I could not tell him. Of course, if his suggestion really had been new and highly valuable, the publication of it with full details would have been of great service to the enemy.

WPB OFFICE OF PRODUCTION RESEARCH

There is another phase of the work of scientists and engineers on developmental matters which also needs to be mentioned. There were broad problems of substitute and strategic materials, and many technical questions involved in the reorientation of industry to the war effort. The development of substitute materials and substitute processes was of great significance as the war proceeded. The task of conducting research and development along these lines was not within the scope of

OSRD, the activities of which were directed to the development of new weapons and their methods of utilization. On the other hand, the War Production Board was deeply concerned with these very matters. In some of its approaches to the problems of materials it was strongly supported by the National Research Council with scientific and technical advice. In November, 1942, WPB established an Office of Production Research and Development under the able direction of Dr. Harvey N. Davis, president of the Stevens Institute of Technology.

ROSTER OF SCIENTIFIC AND SPECIALIZED PERSONNEL

Many auxiliary problems arose in connection with the technical and scientific tasks resulting from the war and prominent among these was the problem of trained personnel. Fortunately, we were able to draw upon the assistance of the Roster of Scientific and Specialized Personnel, directed by Dr. Leonard Carmichael, president of Tufts College, during its busiest years. The Roster and its Committee on Scientific Personnel (Reserved List) rendered excellent service.

NATIONAL ACADEMY OF SCIENCES

This review would be incomplete without mention of the National Academy of Sciences and the National Research Council.

The National Academy of Sciences was formed at the time of the Civil War and operates under Congressional charter. It is charged with the broad duty of advising agencies of government with regard to their scientific and technical problems. Dr. Frank B. Jewett, president of the Academy, has recently described in some detail the enormous burden which has been carried by the Academy and Council in the performance of this obligation.

OSRD leaned on the Academy and Council for scientific and technical advice on many matters. Notably in the medical and metallurgical fields, the organization of NRC provided committees of eminent men who advised continuously and effectively on programs in these fields.

To one who did not work closely with the governmental organization for the conduct of research and development in the war, this rapid survey may seem to indicate a great deal of complexity. It is true that it was complicated, but research and development are themselves necessarily complex. However, the various aspects of the over-all problem were provided for adequately by the organizations that operated in the field and these were tied together as closely as necessary for cooperation by the council of OSRD.

RESULTS OF RESEARCH AND DEVELOPMENT

But what of results? In wartime the work of the laboratory is meaningless unless it finds its way into the field of action. One of the most gratifying expressions of the success of scientific effort in the war is to be found in the attitudes of the Armed Services. Both Secretary of War Henry L. Stimson and Secretary of the Navy Frank Knox indicated their satisfaction over what joint efforts had accomplished in terms of operations. When the story can be told in full, it will be dramatic, and it will reflect the vigorous efforts of a great group of men, employing the best of teamwork in the common cause. It will show without any doubt that the devices developed by American scientists and engineers played an important part in bringing the war to successful conclusion in a shorter time than might otherwise have been the case.

12: THE TEAMWORK OF TECHNICIANS

When I first read of surrendered U-boats proceeding into Allied ports, my mind turned back to 1942 when it was nip and tuck, and the question of whether we could maintain our contact with the United Kingdom, and gradually build up there our forces of attack, revolved about the outcome of a technical race between the U-boat and our means of combating it. The outcome depended very largely on whether there should arrive first on the scene, an increasing U-boat fleet with new devices for eluding attack, remaining submerged for long periods, and with new means of attacking the escorts of convoys; or an enlarged fleet of anti-submarine carriers, long range aircraft, and surface craft for combating them, with means of finding U-boats in the broad ranges of the Atlantic, pursuing them relentlessly whether they remained on the surface or submerged, and attacking and destroying them by powerful methods that could fully overcome their potent ability to fight back. The outcome was quite conclusive and can be simply summarized. The anti-U-boat means and devices arrived first, and the sinkings went down rapidly during the spring of 1943, and were held down by persistence and vigilance to the end. The expected enemy devices also appeared, but too late. The courage, endurance, and sacrifice of the British and Americans on the sea and in the air, armed with the best that science and technology could provide, won a great campaign, one of several, all of which had to be won if our civilization was to survive.

The other side of the record was impressed on me when I read the accounts of the furious struggles on Okinawa, Iwo Jima, and other Japanese-held islands in the Pacific. By reason of far better implements of war, far better skill in their use, and masterly strategy and tactics we successfully assaulted the strongest outposts of the Japanese Empire. In spite of most formidable fortifications and Nipponese fanaticism expressed especially in the use of suicide bombers operating from their home bases, we progressed on such a basis that our total casualties—on land, at sea, and in the air—were far less than the number of the enemy that we had to kill in overcoming their desperate resistance. Our losses were grievous, but the magnitude of the accomplishment is hardly yet realized by the people of this. country.

The struggles did not end on the battlefield. Tens of thousands of American boys were wounded and the effort continued far into the rear areas and the bases, to save life and restore function. The record of the second aspect of the application of science to warfare is magnificent. Medical aid, with the most advanced methods, was available the moment a wounded man dropped. Of the seriously wounded that arrived at base hospitals only a few indeed finally succumbed. These base hospitals were often located in tropical areas that were once disease-ridden and had been largely freed from this scourge. Penicillin, blood plasma, sulphonamides, DDT, atabrine, new treatments of burns, advanced surgery, combined with superb organization and rare devotion to duty to bring to bear every aid that science could offer to heal those who carried the fight directly to the enemy, for us all.

But I cannot recount in detail all of these things. They should be told in due time by those who have directly participated in their accomplishment. I can, however, point out one or two threads that ran through the entire effort, and that should never be lost sight of.

The conduct of this highly technical war produced a great experience in teamwork and professional partnership. Under the stress of war, and in the common interest, men always forget their minor differences, and sink their personal ambitions and inclinations in the heat and intensity of a hazardous effort.

But this has never occurred before, anywhere or during history, to the extent that it occurred this time in the manner in which the United States conducted its enormous war effort all over the world, and it will pay us well to inquire why and how this occurred and what this may augur for the future.

I cannot provide a full analysis here; it will take the lapse of time and the perspective of detachment fully to weigh the factors that have influenced the trends. Moreover, we, who have been close to the center, where military and civilian needs become balanced, where over-complex organization creaks as it tries to adapt, see far more than our share of the disagreements, petty ambitions, and selfish grasping which are on the fringes. The press still finds a story when two men disagree violently in public, and no story when they work in harmony. Those in the field or the laboratory can speak of the solid core on which these minor flaws appear. Their testimony will be that, in spite of all the frailties of human nature, in spite of inevitable cross purposes, there has been a demonstration of collaboration on a scale and to a degree such as has never been seen before. It has been particularly striking in the scientific field.

Military men and scientists arrived at a basis of professional partnership which was extraordinary. Scientists and engineers pulled in the same harness; in fact many scientists became engineers and some engineers became scientists in the process. Scientists from universities and those from commercial laboratories were indistinguishable. Industrialists, government representatives, officers of the Services argued volubly, as red blooded men should, but they pulled in the same direction in the end, and admirable accomplishments were made.

Moreover, the relations between American scientists and those of the British Commonwealth of Nations were just as effective as those within this country, and this is important. Neither group was perfect, of course, but there was just about as much lack of real friction in joint undertakings as in those of a single nationality.

The scientists of this country await the time when they can

pay full tribute to the spirit and accomplishments of the scientists of the United Kingdom. They started earlier than we, and had the longer stress. They often worked under limitation due to shortages of material, and they were bombed. In the early days of the war their vision, and the wise utilization of their efforts in radar and in fighter aircraft, enabled the few to turn back the utmost effort of the Luftwaffe, to save Britain and hence ourselves, and to earn the eternal gratitude of the many. In later years American and British scientists worked so closely together that it will be utterly impossible, and a matter of no vital interest, to attempt to assign many explicit accomplishments to one or the other. In this country we greatly admire British scientists, not only for their eminence in their fields, but for their human qualities. We are proud to have collaborated with them in our joint effort, and we trust that the ties that have been formed will never lessen in strength as we turn toward the advancement of science for the good of humanity in a peaceful world.

I was with Churchill when news from Italy came in, and when General Eisenhower pointed out in his dispatches the great obstacles that had confronted British troops, to offset some of the comparisons of progress that were unfortunately appearing at the time. The Prime Minister quoted these dispatches and said "it warms the cockles of my heart." So too in viewing the interchange and the community of scientific effort on the two sides of the Atlantic, our hearts in this country are warmed, and we look to the future with confidence that the underlying spirit will endure.

But back of this phenomenon, back of all the team work, must lie some great and important causes. I think I have found at least part of the answer, why the two great democracies have given such a heartening exemplification of teamwork, and have thus produced great results.

Recently I have been reviewing the progress of the German scientific and technical effort during the war. The account is still fragmentary, and a great part of it remains confidential, but certain patterns and relations begin to emerge. I have already come to very definite conclusions in my own mind. They are striking, and I believe they will be fully borne out when the full history is written.

If a modern scientific war must be fought, the most effective way in which to fight it is under the temporary rigid controls which a continuing democracy voluntarily imposes upon itself as it girds itself for combat. Such a regime, all other things being equal, can outclass any despotism in bringing to bear on the struggle the combined efforts of science, industry, and military might.

A democracy is efficient in emergency, for the free spirit which it engenders in its normal course is an essential ingredient of great accomplishment under stress. The old contention that only totalitarianism can cope with the complexities of modern life is a fallacy.

De Tocqueville's assertion that the democratic state is less effective than a despotism in a short war, and more effective in a long war, needs to be supplemented under modern conditions of scientific combat. Total complex warfare so emphasizes the advantages of the voluntary collaboration of free men that the democracy will excel in any war, long or short, unless indeed it is so short sighted as to be caught utterly unprepared.

I present these assertions with the conviction that they will ultimately become fully documented and accepted and with the belief that the full appreciation of them is of the first importance in the years ahead. I might go on and state that democracy is the most efficient form even in days of peace with no emergency in the offing, for I believe it, and I hope this too can be demonstrated; but this last point may be academic, for we shall live under the shadow of possible emergency of some sort or other for many years to come.

Germany started her war effort many years before the world became properly alerted to the threat, and it began the development of such things as V-bombs many years before we were similarly at work. In certain areas to which great effort was directed, or in which the Germans had natural aptitude, the technical progress of Germany was great, and at times it even led us in techniques. But these were isolated and relatively unimportant instances. In substantially every important area of the scientific and technical war effort the enemy was outclassed by the great democracies.

The reason? There are many reasons of course. In the latter part of the war the enemy's effort was disrupted by strategic bombing, fortunately for our interests. Slave labor is not efficient labor. But another reason appears to me fundamental. Germany never established partnership, or anything remotely approaching it, between her military men and her scientists. It never brought its scientists, engineers, and industrialists into the common effort with genuine teamwork. Its decisions, on scientific programs, were made in the pattern of all despotic decisions, without the free play and give and take of independent minds, guided by the scientific truth and not by personal fears or ambitions. It badly fumbled the effort at every point. Considering the basis on which that effort was built, one could hardly expect that it would do otherwise.

The German failure was due to many great causes. Its comparative failure in the scientific field was due in no small degree to the fact that true scientific progress, and its effective utilization, prosper well only in the atmosphere of untrammeled scientific freedom. This is only a small part of the great truth that man reaches his peak of accomplishment of mind and intellect only when he is free, but it is an important point, with many practical implications.

With all its enormous advantage of sudden attack upon its peaceloving neighbors, with all its totalitarian regimentation of an entire people for a decade to preparation for the assault, Germany failed. On the scientific front it produced some spectacular and deadly weapons, but it failed in the long arduous race with the great democracies as these applied their scientific accumulations and abilities to their defense. It failed because the atmosphere of freedom is favorable to that collaboration of men of diverse talents which is essential to the effective prosecution of highly complex undertakings. It failed because democracy is the more efficient form; because, when it girds itself for war, it combines the rigid controls which are then essential with the spirit of freedom which it carries over into the emergency. It failed because the regularity and lack of confusion which are the pride of totalitarians are far exceeded in importance in the modern complex world by the effectiveness, by the efficiency, of the untrammeled spirit which develops fully only under freedom.

13: THE QUALITIES OF A PROFESSION

Here I wish to trace briefly the relationship of engineering to the other professions, the professional traditions which engineers inherit, and the outlook for the engineering profession in view of its unique relationship to society. I plan to review the history of professions very sketchily; but through this history runs a thread to which I wish especially to direct attention.

We can start far back, but not tarry long in our review. In every primitive tribe there was some sort of medicine man. He was a man apart, the adviser of the clan rather than its titular leader. He spoke, in his field, with authority, and this rested upon a special knowledge which he was supposed to possess. The medicine man was the progenitor of the professional man of today.

His closest modern counterpart is the scientist. The scientist and the medicine man have much in common. Tribal regalias and feathers have undergone metamorphosis and reappear in cabalistic titles and letters surrounding names. The queer jargon of the cult has changed in nature but preserved its hypnotic effect. Solemn pronouncements about the unknowable still catch the ear of the multitude. The claim to a favored position in society is still based on the occasional ability to unscrew the inscrutable. In fact, one difficulty that faces the scientist is that he may be mistaken for a medicine man, able at will to produce rabbits from hats, instead of the careful, hard-working, human individual he really is.

The descent of the engineer from the medicine man has been highly involved; and it will clarify some obscure relationships if we trace part of it, for there is a central thread which runs through the tale.

The medicine man, and the member of the pagan priesthood which succeeded him, was characterized by numerous attributes. He had a strict code of conduct. He trained neophytes, subjected them to a long period of apprenticeship, initiated them into the mysteries, and inculcated in them pride in the cult, and rigid discipline in its formulas. He severely restricted his numbers, by intellectual hurdles to be surmounted. He spoke a special language. He sat as adviser in councils of the mighty. But, more essential than all of these, he ministered to the people.

This was the first professional group, and all others have derived from it. Not every attribute has been maintained as new professions have emerged; but to a surprising extent their counterparts can still be found. In every one of the professional groups, however, will be found the initial central theme intact—they minister to the people. Otherwise they no longer endure as professional groups.

Ministry needs definition for our purposes. The alteration of word meanings with new usages is such that it is only too easy to be misunderstood. Ministry is not service, and we have so completely altered the essential significance of the latter word that it may have utterly different connotations to different hearers. Ministry carries with it the ideas of dignity and authority; it connotes no weakness, and offers no apology. The word has been carried into diplomatic usage; and in the derived form of *administer* into law and business. There is no fog of subservience surrounding the concept. The physician who ministers to his client takes charge by right of superior specialized knowledge of a highly personal aspect of the affairs of the individual. The attorney assumes professional responsibility for guiding the legal acts of his client, and speaks with the whole authority of the statutes as a background. It is in this higher sense that we trace the thread of ministry to the people.

This is the fuel which has kept alight through many ages the professional spirit. Every time that the fuel has become exhausted, the light has gone out. It has not mattered how much was retained of trappings and mysticism, nor what the profundity of utterances, there has been no true profession that has not with dignity and authority advised and counseled the people, that has not guarded the commonweal. For a true profession exists only as the people allow it to maintain its prerogatives by reason of confidence in its integrity and belief in its general beneficence.

The monastic orders, under divers religions, springing up as outgrowths of the simpler systems of priestcraft, have exemplified the theme in two ways. Some have preserved, adorned, and extended the knowledge of their time and place. These have their modern counterparts in the scientific and learned groups, the custodians of our culture, and the source from which flows new knowledge for the use of man. Other orders carried to great heights the direct ministry to those in misfortune or distress—often at great self-sacrifice, as did the early Jesuits among the Indians of our west. Both forms have remained high in the esteem of the people and have endured. Occasional groups have lost the thread, and have, for example, become militant orders devoted to self-aggrandizement; and these have disappeared.

THE TEACHING PROFESSION

Out of the early priesthood came also the teaching orders, whose ministry took the form of instruction of the young, and this aspect of professional activity is represented today by the great profession of teachers everywhere. This group has little indeed of the trappings of the medicine man, it has no single closely knit aggressive society representing it; its language is becoming complicated but is still fairly intelligible to the layman. Where it has maintained its ideals it is honored and respected. Great teachers do not find riches heaped upon them, they do not become affluent. Great teachers have no interest in riches. In the great teacher the parental instinct, which is so often at the basis of senseless extremes of individual striving for wealth, becomes sublimated into a broad love of youth which calls for neither wealth nor power for its enjoyment and satisfaction.

THE MEDICAL PROFESSION

A very early offshoot was the profession of medicine, for ministry to the ill was a primitive need. It has had a long and distinguished history. Utilizing the fruits of science it is today in full tide of accomplishment for the benefit of mankind. It has had vast power and influence. Yet, today, in the United States, it is at a turn of the road, and its most thoughtful members are giving earnest consideration to its future. There is serious danger that its light may fail, and its heritage of idealism may be lost.

The profession of medicine, by reason of its very nature, has preserved many of the attributes of the ancient forms. It selects its neophytes by rigorous intellectual elimination, trains them over many years, and seeks to endow them with the philosophy of their profession. It severely restricts its own numbers, perhaps too severely in view of the task before it. It preserves itself apart, by special language, and has a unique code of conduct. It has sat in the councils of government and advised. By the will of the people it has been given special privileges and prerogatives for use in pursuit of its objectives. It is highly organized.

Through long ages it has held well to its ideal of simple ministry to the people, and has disciplined under its codes those who would use its special privileges for other ends. It has guarded the people against their own folly, and has been properly militant in maintenance of its sphere of the public weal. Its individual members are in general respected in their own communities to an extraordinary degree.

Yet, in these days when all institutions are undergoing scrutiny, when our population in fear and distress is prone to be critical, there is evidence about us that the profession, as a profession, does not command that full support of the people of the country without which it cannot continue on the path. Yet as one looks about in the medical profession signs are seen of a resurgence of idealism, a re-emphasis on the simple mission of healing, and a recognition of the central theme of ministry to the people. This I am convinced is the true motive of the great majority of the members of this grand profession. Yet there is much suspicion in the public mind that aggrandizement, utilization of power for the professional advancement of the membership, the guild spirit in its cruder form, are rampant. I second the thought of many eminent members of the profession itself that unless this suspicion is allayed by a revival of simple ideals, the profession will suffer, and the people will suffer enormously with it. It is well that engineers should be deeply interested in the outcome, for medicine is a very ancient profession from which we have much to learn.

THE LEGAL PROFESSION

To treat the origins of the profession of law, its codes and countercurrents, would require an article in itself. Here is a field in which the preservation of the true philosophy of a profession is intricate indeed. Endowed with special privilege under the law, it largely regulates its own conduct. Never quite successful in the recruiting, training, and indoctrination of its neophytes, its maintenance of adherence to a high code of conduct is rendered more difficult. Counseling with government and, by the nature of its mission, participating directly therein, it has great power for good or evil. It certainly strives, as an organized profession, for the public welfare; but its zeal in this regard is not always such as to cause it to disregard the special welfare of its own group; and the two, withal, are sometimes hard to disentangle. It ministers to those in legal distress with great effectiveness; but the distressed often appear in pairs. It is hardly judged as a whole by the public. Certain it is, however, that those of its membership, on the bench or at the bar, who have risen to the highest positions in their devotion to professional ideals, are respected and honored by the public. Certain it is also, that, should this respect falter, we as a democracy would soon be in a sorry state.

THE ENGINEERING PROFESSION

But our principal concern here is the engineering profession, and we inquire, what is the engineering profession; is it a profession at all; and if it is, will it develop into the full stature to which the importance of its works entitles it to aspire?

It is relatively young. The military engineer appeared in the first steps of the mechanization of warfare, when forts began to take shape. His counterpart in peaceful affairs was called a civil engineer. With the industrial revolution, and especially with the spread of mechanization from the factory into every walk of life, engineering became exceedingly diversified. Applying science in an economic manner to the needs of mankind is its broad field. Its disciplines are spread over all the sciences as they become thus applied, and embrace also portions of economics, law, and business practice which are integral parts of the process of application. It is somewhat loosely organized as professions go. To a minor extent only, it limits its numbers; but the very strictness of its essential disciplines provides some selection of its neophytes. Until recently it has done very little in an organized fashion to inculcate in its younger members the philosophy of the profession, leaving this largely to those of its individuals who are also members of the teaching profession. That branch which represents the consultant, and others to a degree, has drawn codes; but there is no body of codified principles which is accepted and applied by the profession as a whole. It has no highly distinct language or jargon, for it must continuously work with laymen.

These are, however, incidentals. The important point is this: Does it have a central theme of ministering to the people? Most certainly it serves the public in myriad ways, but are its individual members activated primarily by the professional spirit of dignified and authoritative counsel and guidance?

THE BUSINESS PROFESSION

In order properly to inquire into this weighty question, we need to digress a moment to consider another large group of the population, the modern men of business who have derived from the ancient traders and merchants. The merchant class has not usually been a professional grouping in the true sense; and engineering, which has derived its philosophy from this source as well as from science, naturally partakes of the heritage of both groups. Business has served the public, of course, but its main theme has been the profit motive, a salutary objective when restricted by law to the use of ethical procedures in its pursuit, but not a professional objective.

One of the most encouraging signs of the times is the gradual emergence in our day of the truly professional man of business. Scattered, not organized, with no sign of professional trappings, they are nonetheless possessed of a high mission, which needs only formulation and recognition in order that they may constitute a new and strong profession. This is occurring for one reason because of a gradual change in corporate form. The owner-manager was activated by the profit motive, and no amount of paternalism could wholly alter his position in the social scheme. Even with corporations, the ownership of which is widely scattered, the manager is ordinarily controlled by and primarily responsible to a few powerful owners, so that he in essence still represents the interest of the owners in his relations with the three bodies with which he deals: the government, the employees, and the consuming public. It is his difficult task to reap for the owners' benefit the fruits of his industrial operation, while maintaining at least tolerance on the part of the other bodies. But there are some corporations in which ownership is so diffused that the management becomes in effect a self-perpetuating entity, partaking therefore of the nature of a trusteeship, with equally weighed obligations and responsibilities to all four bodies: owners, employees, government, and consumers. Among such managing groups will be found individuals who have the professional philosophy in high degree, conducting their affairs for the just and equitable benefit of all four groups concerned, maintaining the health and progress of their institutions as potent agencies for ministration to the needs of the people. They find common ground with the trustees of great foundations, of hospitals, and all non-profit organizations devoted to the public welfare. They find common ground also with many of those who make a career of the business of government. Their ranks are recruited by many in the ordinary walks of business who have seen a light and envisioned a function in life which is higher in its satisfactions than the struggle of any body against any other; namely, a struggle with all bodies to preserve an ideal. Out of this trend, as competition for industrial existence becomes tempered, should emerge a new profession with its own traditions and beliefs, which is capable of managing prosperity so that it will be conducive to the health of a nation; and there is grave question whether this objective can be attained in any other way. I wish there were a special order of knighthood in this country to honor and unite those who are now blazing the difficult path and developing the novel philosophy of this new profession.

Engineering, however, derived jointly from the quiet cloisters of science and from the turmoil and strife of aggressive business, and it is no wonder, therefore, that it should wobble a bit as it seeks to evolve its own professional philosophy. Just as it is not reasonable to expect the young neophyte to grasp at once the idealism of his calling, so it is perhaps not reasonable to expect a profession which is so young and which has grown so fast to have found itself in this regard.

ENTRANCE INTO A PROFESSION

The period of initiation into any profession should extend into maturity. Only when members reached the full bloom of manhood did the ancient orders entrust the mysteries to their care. The young neophyte served his apprenticeship under constant tutelage and close guidance by mature minds, and this we still find in every profession. As apprentice, as employee, he is called upon to prove himself before he enters into that relationship where his opinions are controlling in his special field, and some there are who never emerge from close control and the mere exercise of technical proficiency. In the engineering profession this emergence usually is circumscribed by the fact that most engineers operate as members of industrial organizations of one sort and another, and the fact that they serve their apprenticeship in this same sort of organization and come to devote their entire efforts to its affairs, rather than to enter them after professional recognition elsewhere, as is usually the case with medical or legal individuals. This, however, merely emphasizes the need for better supervision of the neophytes by the members of the engineering profession who have arrived. It is not enough to leave their training to the industrial organizations of which they are junior members. Inculcation of the principles of the profession can come only from those who themselves have attained to a full grasp of its proper function in society, who have arrived at a balanced judgment as to its responsibility to the several groupings of which society is composed, and who have a professional interest in the young men who are destined to succeed them in the profession. Every profession should have its secrets and its mysteries spread before the world that all may read, but truly grasped only by those who have lived the professional life; and these should be transmitted to the neophytes with due care, with reverence for their inherent worth, and in due time. Ritual and symbolism, secrecy and circumspection, were the ancient paraphernalia which insured a proper seriousness in youth in order that the impartation might be impressive. These have not wholly disappeared from modern professions. Admission to the bar, the use of the title of doctor, and similar customs and usages have profound effect in producing a professional consciousness. The engineering profession is wholly without these aids, and its task of inducting its neophytes into the true professional atmosphere is thus rendered doubly difficult.

But does it matter after all? Are the things that engineers do so vital that they must needs be approached in the professional spirit? Most certainly it matters. And most certainly the task is a professional one. The impact of science is making a new world, and the engineer is in the forefront of the remaking. He lights the way in a very literal sense. He brings peoples close together for better or worse, by facile communication and rapid transportation. He guards the food supply, and replaces the hopelessness of Malthus with an embarrassing plenty. He shortens the hours of labor, and fills the consequent leisure with distractions. He temporarily disrupts the techniques of whole industries, and thus alters the life habits of many people, in maintaining a continually rising standard of living. He bores through the earth and under the sea, and flies above the clouds. He builds great cities, and builds also the means whereby they may be destroyed. Certainly there was never a profession that more truly needed the professional spirit, if the welfare of man is to be preserved.

EXPRESSION OF PROFESSIONAL SPIRIT

There is no lack of signs of a rising consciousness in this regard. The profession is most positively vocal. There is a vigorous new organization, linking several large groups, devoting itself to the improvement of the education of the young engineer, and the instillation of high principles during his early career. Engineering literature is full of discussions of the duties and responsibilities of the profession, and out of this may crystallize some day a code, a set of principles of conduct, a guide drawn solely with the object of advancing the public weal, which will become accepted by engineers everywhere, whether in government employ, private practice, or industrial organization. Having, to some extent at least, consolidated their techniques, engineers are certainly giving thought and voice to their position in society, and to their responsibility for the use of the great works which they create.

The focus of this whole affair is the American Engineering Council. More than any other group it represents the engineering profession as a whole, in its relationships with government, other professions, and the public. Here, more than in any other organization, reside the external as contrasted with the internal relationships of the profession. It was founded by men who considered its functions in terms of a high idealism. It is now going through a strenuous period of self-examination. To this every individual can contribute only one set of thoughts to be merged with all of those which seethe, and interact, out of which will come in due time that consensus which will form the opinions, traditions, codes, and consciousness which will mold the engineering profession. It will come unless the Council fails; for if it fails, and if its place is not taken by a more rugged successor, there will be no unitary engineering profession at all. In the spirit of adding my few thoughts to those of the eminent men who are directing the Council I have previously offered comments, and I now comment again, with the expectation that I will be disagreed with and answered, with the wish to add my mite to the consummation.

I find it a vigorous and rapidly evolving body. I consider it to be utterly inadequately supported by the profession as a whole, in comparison with the central bodies of sister professions, and with a serious problem as to how adequate support can be drawn for the great task that lies ahead of it. I find it partaking of the great American tendency toward overcomplication, and inclined to attempt things which seem to me personally to be off the main beat. I find to my great joy that it is gradually becoming known and recognized; and I trust this is just a beginning. I find it guided by some of the best minds in the profession as its officers, who are giving valuable time to its cause; and I hence cannot fail to be optimistic as to its future.

To me, however, there is just one point on which I wish to focus attention. I find it struggling with its own philosophy. I find, in fact, that it hesitates as it formulates its idealism; that it has not yet placed its foot unequivocally and irrevocably upon the path that leads to complete devotion to the public welfare. I find that it has not yet enunciated its belief that the great mission of the engineer lies in intelligent, aggressive, devoted ministration to the people. This I urge with all the emphasis I can command.

NECESSITY FOR A CENTRAL ORGANIZATION

I do not seek to conjure away practical difficulties by ignoring them. I know full well what restricted budgets mean. The argument that the support of the membership can be obtained only if they can see a direct and personal benefit from their contributions has a familiar ring. I recognize that it is entirely proper for professional men to join in an insistence upon a reasonable and proper recognition of their services to society. Yet if there is no central organization which has as its creed the best service of the profession to the society of which it forms a part, then there will be in the end no engineering profession. Professional status rests in perpetuity, not on transient law, not on the cruder machinations of the ancient guilds, not on exclusive control of those having a specialized and necessary knowledge; but upon the respect and fundamental support of the people who are served, who only in the long run can insist upon the maintenance of prerogatives, and confer honor, recognition, and special privileges in society upon the members of a profession.

Will engineers support such a program? Will they contribute directly or through their specialized societies to the development of this ideal, and its exemplification in Council projects aimed at rendering real some aspect of the profession's contribution to public welfare? Will they make possible great forums for the crystallization of engineering opinion on public questions involving engineering, not to attain an impossible unanimity or produce high-sounding resolutions, but so that all aspects of controversial matters may be aired in order that people may know what engineers think? Will engineers go along heartily in developing a professional consciousness, a code of action, a philosophy which implements a desire to be a truly professional group, oriented primarily toward the advancement of the public health, safety, comfort, and progress? Will they accept as the central theme the engineers' ministration to society, without fear of any class, and without prejudice toward or away from any special social interests or causes?

If they will not, then there is no truly professional spirit to build upon. We may as well resign ourselves to a gradual absorption as controlled employees, and to the disappearance of our independence. We may as well conclude that we are merely one more group of the population, trained with a special skill, maintaining our economic status by a continuing struggle against the interests of other groups, forced in this direction and that by the conflict between the great forces of a civilized community, with no higher ideals than to serve as directed, and with no greater satisfaction than the securing of an adequate income as one member in the struggle for the profits of an industrial age.

But I know the minds of too many engineers to be thus pessimistic. I recognize the distinguished careers of a generation of men who have practiced in the profession to its credit and honor. Though the task be difficult, on account of the nature of many of our relationships to society, nevertheless traditions are being formed, the consciousness of the membership is becoming aroused, and I confidently expect the profession of engineering to develop in a manner of which we can be justifiably proud.

And to those in the ranks who may not have yet seen the light, I would preach the doctrine, without pulling any punches, without mincing any words, that the path of professional attainment lies open before them, that it is a thorny path that is easily lost sight of among the rocks and rubbish, that it can be adhered to only by sacrifice and by support of those who lead the way, that it is a long path which leads down into the valley into which the sun does not shine, but that it leads at last to the heights—to the heights of true professional attainment, where honor and individual recognition by fellows is the real reward, and where the watchword is that old, old theme, which has never lost its power, and which may yet save a sorry world, simple ministration to the people.

14: OUR TRADITION OF OPPORTUNITY

Thomas A. Edison exemplified an important aspect of American life. He combined, in superlative degree, resourcefulness and initiative with an intense practicality and a keen vision. The many results which he attained during his life were the blossoming of this combination of talent in an environment which was wonderfully adapted for such a growth.

I do not need to say that these are not the only characteristics that are needed in scientific and engineering endeavor. The list of Edison medalists is itself a demonstration of the recognition of the need for combining such qualities with scientific deduction, mathematical analysis, and the like, in order that there may be a rounded whole. Our great industrial advance, with its intense use of electric power and its intricate mechanization, has arisen by reason of the efforts of many types of mind. But without the important phase that Edison so notably exemplified the scene would indeed be a drab one.

The United States has prospered in a material way. It has attained a standard of living far beyond that arrived at in any other country. This is not entirely explained simply by the presence of great resources, or a large and uniform market, or even by the opportunities for pioneering that are summarized in the convenient term "the frontier." Neither does the opportunity for advancement necessarily vanish when the geographical frontier gives place to one that is entirely of a technical nature. The advancement has occurred because America brought a new idea into the world—a valuable idea.

In a word, there was produced an atmosphere in which Edison could function, and in which men like him in having ideas and the intellect for their furtherance, however much they might differ from him as to method, could make their influence felt. The startling forward steps that Edison caused by his own efforts could not have occurred in a totalitarian state, whether the label were that of state socialism or Fascism or something similar. They could occur fully only under the unusual state of circumstances which obtained in this country when Edison worked and created. The atmosphere that existed is hard to define; it had its crudities, but it was a unique atmosphere and it is well worth preserving.

As the United States matures, we are in distinct danger of losing this enormous asset. The unwillingness to take risks which accompanies economic maturity, the cross-currents of pressure interests, the mere increase of interdependence due to the advance of mechanization, all tend to destroy the flexibility and freedom of action which are essential parts of the atmosphere of individual creation.

To the trends that were molding our environment and conditioning the atmosphere within which our pioneers worked was suddenly added the impact of war. We awoke late from a period of lethargy to find ourselves in an intense struggle caused by the clash of our political and social philosophies with ideologies far different from our own. We shall not return to the exact condition from which we departed when war came. We have learned many things, and it is to be hoped that we shall not forget them. It is of especial importance that, as we readapt ourselves to a new period of peace, we should preserve the essential features which made this country great. In particular, the opportunity for the individual creator, for the industrial pioneer, for the inventor with vision and practicality should not be lost; and the atmosphere of our industrial life should be favorable for his efforts in our behalf.

The effort to hold our own in a technical contest was not by any means a sinecure, or a struggle in which the outcome could be predicted a priori and with full confidence. In the course of the war with the Axis nations we had to mobilize our technical and scientific resources to the full, and we had to do so with the confusion that is inevitable when a great lumbering democracy suddenly turns to war. It was necessary to regulate, to impose controls, and to operate with a secrecy that is foreign to our own usual open methods. It was not necessary, however, to suppress thought and initiative and the democratic functioning of scientific and technical groups devoted to important phases of the technical aspects of the war effort. The results were good.

When the full tale is told, or the part of it is told which properly can be realized, I am convinced that it will show one thing and that it will show this forcefully. The free untrammeled science and engineering of a democracy, when it once becomes directed toward an objective with full vigor, can outstrip the regimented efforts of any totalitarian state, provided there is anything approaching equality of resources on the two sides. I believe that it will come to be realized fully that our progress in manufacturing with enormous speed the weapons needed by our Armed Forces, in improving these weapons, in initiating new ones, and in keeping our civilian economy with its enormous technical complication still running effectively, was not matched by the enemy, and that this occurred in no small degree because our wartime effort was built upon the existing structure of freedom, with all that the word implies.

This must be preserved. As we turn again to the days of peace, it must be preserved in spite of the trends that flow from the greater interdependence of society, in spite of the trends that are inherent in the growing maturity of a nation, in spite of selfish interests of every sort. There must remain in the United States the opportunity for an Edison, the opportunity for any youth with initiative, resourcefulness, practicality, and vision, to create in his own name and by his own efforts new things that will tend to make this country vigorous and strong and safe.

The way in which this can be accomplished is not immediately apparent. The difficulty does not reside in the attitude of the people, for certainly there is an overwhelming majority that holds strongly to the conviction that freedom of individual opportunity must not be allowed to lapse, and that this involves genuine industrial opportunity for the individual, or the small group of individuals that join in a business effort. The difficulty arises from two facts. First, there are great sections of our industrial affairs that can be handled economically only by large industrial units. Second, the very complexity of modern life requires increased and centralized governmental activity, in order that the public interest may be fully protected and furthered by those measures of regulation and public works which government alone can perform adequately. The point is that, as a people, we have two parallel objectives; and we have been clumsy at times and allowed one to submerge the other.

OUTLETS FOR INITIATIVE AND TALENT

In time of war concentration of effort and the imposition of rigid controls are essential for a fully coordinated all-out effort. The question is raised why, if this concentration is more effective for war, is it not also more effective for peace? One answer is the fact that the war concentration is effective primarily because it was constructed out of elements that became available and efficient under the relative freedom of peacetime conditions in a democracy, and there is no assurance that a brief effective concentration would remain so. Quite the contrary is what I believe the record of history will show. A more complete reply lies in the fact that we seek in the United States something beyond mere mechanical efficiency; we seek a society in which initiative and talent may have an outlet, and in which the individual may have opportunity to rise by his own efforts and contributions and not merely by the fixed operation of a system. We would be willing to sacrifice, if need be, some mechanical efficiency in the interests of individual freedom. My own conviction is that no sacrifice in the full effectiveness of the country is truly involved in the long run; and that, on the contrary, the elements which have rendered us strong in the past will render us still stronger in the future if we have sufficient intelligence and conviction to insist upon their preservation.

As peace returns, controls automatically become relaxed. This is not enough. A negative slump back to a state of drifting, buffered by the trends that are inherent in technical progress, will not carry us on our appointed course as a nation. A positive and well-thought-out effort is needed if we are to combine the conditions of modern industry and modern government with that freedom within unitary organizations and within society as a whole which will allow the unusual individual to have a real chance of accomplishment and success. It can be done, if we have the steadfastness and conviction to insist that it be done. In this effort the great body of engineers and scientific men in the United States have an important part, for they are in a unique position to view and understand both sides of a matter which is primarily technical, economic, and organizational. If we are wise there will be, in the future, many Edisons in the United States. They may not shine with his peculiar brilliance, but they will add to the well-being of the nation a necessary element which can be added in no other way. Their opportunity must be preserved open before them.

15: THE NEED FOR PATENT REFORMS

The patent system of the United States was set up originally to bring benefit to the public by advancing the useful arts. It does so by creating a temporary monopoly, thereby rendering possible the hazardous development of untried inventions, which would otherwise not come to fruition to add to the general well-being and increase the standard of living of the people. By its substantial rewards it stimulates invention, and the assiduous study and persistent effort on which invention is based. That it has been successful needs no demonstration, for its results are all about us.

The primary purpose of the patent system of this country is to stimulate new industries. This is always an important matter, but it becomes particularly important as the country emerges from a serious depression. The history of depressions shows that the time of emergence is usually marked by important technical advances resulting in the creation of new and extensive industries. If this had not occurred we could not have attained the present high standard of living. For the prosperity of the country it is imperative that this trend should continue.

The patent system in the past has been one of the primary influences in shaping American industrial life, and it has assisted enormously in the development of the country. In the considered opinion of those best able to judge, it is not however at the present time functioning to full advantage. There are serious difficulties. The use of scientific results in industry is a much more complicated matter than when the patent system was first set up, and the system has not been altered to bring it closely in line with the modern complex matters with which it

151

has to deal.   If it is to fulfill its proper function to the greatest possible extent it is therefore essential that it be changed in certain ways in order that new industries may be stimulated and not inhibited by its operation.

The patent system of this country is old, and it has gradually developed into a complex structure.   Radical changes in such a system should of course not be undertaken without serious and careful consideration.   It would be equally fatal, however, to refuse to consider alterations at all when the changed times dictate modification.

In a complicated situation such as this, it is not possible to point out panaceas which will automatically treat for the optimum public benefit and with complete equity every individual case that can be cited.   Objections can be raised, and will be raised, to every suggested change in a system which so closely affects the interests of widely different classes of individuals. The attempt has been made to recommend as few changes as possible, and to make these changes in such manner as to bring the greatest good to the greatest number.

There are three primary defects in the system as it stands at present, considered in connection with the functions which it is called upon to perform in a modern complex technical world. The first defect arises by reason of the issuance by the Patent Office of an enormous number of patents, many of which should never be issued, chiefly because of an unduly low standard of invention.   The second defect has to do with the excessive cost and delay in the litigation of patents, by reason of the present system of appeals.   The third results from the difficulty met by the courts in handling scientific or technical questions without competent nonpartisan assistance.

As these defects exist there is a question in the minds of many serious minded and experienced men whether the system is not after all more of a liability than an asset.   It is seriously suggested that the system has become so complex and cumbersome that it may break down of its own weight.   The situation, while serious, is not at all hopeless; it is possible to make certain changes in procedure, not in themselves difficult to put into effect nor expensive and not changing the existing structure in any essential or radical manner, which yet may restore the system to its former condition of importance and beneficent influence on American industry.

PRESUMPTION OF VALIDITY

The Patent Office now issues many patents which are later found invalid in the courts. It issues a much larger number which never can have commercial importance. With two million United States patents issued, the situation is unduly complex and is growing worse. When approximately 90,000 patents are applied for in a year the amount of attention which can be paid to each one in the Office is not sufficient to insure a strong presumption of validity in issued patents. The staff is overburdened. It has neither the opportunity nor the facilities to make the study and search necessary to clarify the situation, and the trivial and the obvious are issued to confuse American business. This situation is not the fault of the Patent Office personnel. It results from the nature of the technical advance which has taken place in the past few decades. It should, however, be positively corrected.

The standard of invention cannot be arbitrarily raised by creating a new definition of invention. The courts can influence the standard through their decisions only gradually, and by the undesirable means of finding invalid a large fraction of the patents which come before them, which, temporarily at least, decreases rather than increases the presumption of validity of patents as issued. The Commissioner of Patents should be supported in his efforts to eliminate the trivial and the obvious. but merely increasing the number of patent office personnel will not effect a cure. There is needed a change in procedure which will aid the office in raising standards, and positively increase the presumption of validity.

PUBLICATION BEFORE ISSUANCE

The British and German systems provide for publication of an application before issuance, thus inviting contests within the Office prior to the issuance of a patent. There are obvious objections to this procedure. The most serious objection is that the inventor is often unduly burdened with the expense of a contest, which is particularly serious for the individual inventor without resources. It is much better procedure to maintain the action in the Patent Office ex parte, as is our present practice. However, without incurring the difficulty of the system involving contests, it is possible to secure much of the improvement in the presumption of validity of an issued patent which such a system produces. This benefit is very real. At the present time our Office issues patents without a thorough search of American and foreign literature, but with a search often devoted to American patents only, with some small attention to publications and foreign patents. The result is that many patents are issued which are clearly invalid in view of prior patents and publications. Such patents often cause expensive litigation before they are finally found invalid. The theory that the Office should issue patents with little or no examination, leaving the determination of their validity to the courts, is either practically inoperative or unduly expensive. This is substantially the French system. The American system is preferable, and it goes a certain distance toward the examination of prior art in order that a patent when issued may carry strong presumption of validity, instead of being merely a means for entering litigation. However, our procedure does not go far enough, and the provision of an adequate corps of examiners, with sufficient time and training to be able to review adequately the entire prior art, whether in patents or in the literature, would be highly expensive. A modification of the system of publication before issuance will secure the desired result without great cost. It will aid the Patent Office in increasing the presumption of validity of issued patents. It is recommended, therefore, that, when an application is ready for allowance, it be published in the Official Gazette, and the submission of pertinent facts by interested parties be invited.

The publication should be made in the manner employed at present in publishing an abstract and sample claims when a patent is issued; and the allowed claims, and preferably also the specification and drawings, should be opened to inspection. The publication and material opened to inspection should not disclose the date of filing, nor give any other information unnecessary for the purpose in hand. Upon such publication the Office should allow anyone interested, and within a stated time, to submit facts which are pertinent to any application thus published. These facts, however, should be limited to references or photostat copies of prior patents or other published printed papers, books, or documents, such as are available in libraries or other public sources.

Arguments and affidavits should be rigidly excluded. The procedure in the Patent Office should be maintained strictly ex parte. However, before the patent is finally passed to issue, the Examiner should give it a further review in view of any new material thus brought to light, and either pass it to issue, or make necessary rejection of claims. Of course in case of rejection on this basis the applicant should have an opportunity to present arguments as he has at present, and an opportunity of appeal. The documents filed should be made part of the file-wrapper of the application.

An applicant who files an interfering application after such publication should be under the same heavy burden of proof as the applicant who now files an interfering patent application after the granting of a patent.

This change will not cause undue expense to the inventor, but will aid him by giving him a stronger patent, much less likely to be voided by the courts. The burden of submitting evidence will be welcomed by those interested in special fields of development, as it will largely avoid the more serious burden incident to the issuance of unwarranted patents.

It appears that this change can be effected by amending the Patent Office Rules of Practice. A relatively small increase in expense of operation of the Patent Office is involved, and this should be provided for in the proper Congressional legislation.

A great deal of delay and confusion results from our present system of litigation of patents. The patent suits on a single important patent may cost several hundred thousand dollars. Such a burden confronting a young and struggling new industry often results in its thorough discouragement. It is possible under the present system for very many years to elapse between the initiation of proceedings and their final disposition, and industry in the meantime falters. It is possible for suits to be brought simultaneously on the same patent in several district courts. Moreover, on their appeal to the circuit courts of appeals it is sometimes the case that conflicting decisions are given in different circuits. The result of this entire situation is a serious burden on growing industry, and on this point there is the strongest feeling among users of the system of a need for simplification.

A SINGLE COURT FOR PATENT APPEALS

There should be established a single Court for Patent Appeals, in order to establish and maintain harmony and accuracy in judicial interpretation of patent questions, by confining the appellate jurisdiction in civil patent causes to one court, composed of permanent judges having the necessary scientific or technical background.

Each judge should be learned in the law and proficient in knowledge of the industrial application of science, and should have had a reasonable experience in the trial of patent suits on the bench or at the bar. If, in order to grasp more fully special technical questions, the court wishes to call temporarily upon experts to advise and consult on difficult points, it should be enabled to do so.

In view of the importance of this court the salaries paid to the judges should be adequate to attract men of the highest stamp. The qualifications have two aspects, and it is accordingly desirable that scientific as well as legal opinions and suggestions concerning appointees be given weight.

In the phrase "civil patent causes" we include suits in Federal Courts, other than the Court of Claims, (1) alleging infringement of a patent, (2) alleging breach of a license agreement involving a patent or invention, (3) in equity to obtain a patent, (4) in equity alleging interfering patents, or (5) under the declaratory-judgment law, involving any of the above issues.

The Court should be composed of a sufficient number of permanent judges, any three of whom should constitute a quorum. The Court should be located in Washington, D. C., and should also hold terms at least once a year in each judicial circuit, except as these may be omitted at the discretion of the senior or chief justice of the Court.

It appears desirable that there should be transferred to this new Court the present jurisdiction of the Court of Customs and Patent Appeals of all patent and trade-mark appeals from the Patent Office. Emphasis should be placed on the desirability of a single court, adequately provided for, composed of judges of high qualifications, with final jurisdiction in patent causes except as their findings may be reviewed by the Supreme Court on writ of certiorari. Such a court will bring to industry that certainty and expedition which is essential if the patent system is to be fully effective in stimulating new industries.

In order to put this recommendation into effect Congressional legislation is needed.

ADEQUATE ASSISTANCE TO COURTS

The determination of the just equity in a patent suit involves two diverse aspects, the law and the technical facts. When the technique involved was simple, before science had made the great strides of the past generation and before the fruits of its progress became applied and embodied in patents, the judge could readily acquire during the progress of a suit that background necessary for him to understand the technical facts presented to him. To expect him to do so today, with the present specialization and intensification of technical knowledge, leads to a severe burden upon him, and to undue expense to the litigants. It is true that the litigants call their own experts, but this does not fill the need. The Court itself should be so composed as to understand and deal adequately and promptly with the matters brought before it. This has been embodied in the previous recommendation of a single court for patent appeals. It is especially desirable that courts of first instance be also so constituted as to treat difficult technical questions with precision and promptitude.

ADVISERS TO THE COURT AND TECHNICAL JURORS

It is therefore recommended that there be provided scientific or technical advisers or juries to furnish adequate scientific or technical assistance to courts of first instance in equity patent causes.

The phrase "equity patent causes" is used to exclude suits at law, but is otherwise synonymous with "civil patent causes" as used in the preceding section.

The advisers or jurors should be United States citizens of sufficient scientific or technical qualifications so that they are expert in the art to which the suit relates. They should be selected by the Court, with such suggestions from the litigants as may be solicited, but without the necessity of securing agreement of the litigants to the selection.

Initially they should be selected at large. It is recommended however, that steps be taken to prepare and maintain an adequate list of qualified experts, and that upon its establishment selection should be confined to this list. It is believed that the National Research Council, in cooperation with the national scientific and engineering societies, would be the proper agency to be charged with the duty of preparing and maintaining a list for this purpose.

It has been stated that it would be difficult to find properly qualified experts.   No such difficulty will exist.   It is true that there are many fully qualified scientists and engineers who consistently decline to act as experts for litigants in patent cases; often because the partisan presentation of a cause, while necessary and proper, is natural for an attorney but unnatural for a scientist or engineer.   To a call from the courts for dignified and non-partisan aid in the handling of patent cases there will be ample response.   Nor does this country lack men of the highest type, both from the standpoint of their professional attainments in the sciences and their applications, and from the standpoint of their trustworthiness and public spirit.

It should be mandatory upon the Federal district courts in equity patent causes to utilize the services of either a technical adviser or a technical jury, but the court should be free to select either alternative, and should make selection anew for each suit.

When a technical jury is utilized, its report should be final as to questions of fact.   Three jurors should be sufficient.

When an adviser is utilized, he should be merely advisory to the court, and his report, if called for by the Court, should have the same presumption of accuracy as a master's report has, under the Equity Rules.

The adviser or jury should act in conjunction with the court and under its direction as to procedure.

The compensation of experts employed in this manner should be commensurate with their usual earning power.   It should preferably be fixed by the court, as is done now with masters under the Equity Rules, but it may be fixed by statute, in which event the maximum per diem should be such as is customary for consultants with high standing in their professions. This compensation may be taxable as part of the costs of the suit, as is done now with masters under the Equity Rules; or it may be paid by the government as a part of the cost of maintaining the courts.   On the matter of the allocation of the expense no convictions are offered.

This modification in procedure will notably and properly increase the prestige and dignity of the courts. It will utilize, in the speedy and just disposition of patent causes, the great asset which this country has in its body of scientific and technical men. It will, by causing expedition, decrease the costs of litigation; and by rendering our patent system more sure and effective, it will benefit especially inventors and new industries, and thus benefit the people generally.

It appears that this change can be largely effected by the United States Supreme Court through an amendment to the Equity Rules, although Congressional legislation may be needed on some points.

COMPULSORY LICENSING

There have been repeated suggestions that some system of compulsory licensing be introduced in this country. The usual reason given for the need of such a system is that patented articles are sometimes not manufactured and made available to the public, for one reason because of the failure to reach an agreement on the part of those owning several patents, all of which are involved. The principal argument against compulsory licensing is the statement that by decreasing the strength of the patent monopoly it would reduce the incentive to invention and development, and vitiate to a considerable extent the effectiveness of the system in the development of industry. The point is a difficult one, and it goes directly to the heart of the system.

No system of compulsory licensing should be introduced at this time. This problem has not as yet been constructively analyzed with the completeness which should precede any such fundamental alteration in our patent system as is here involved. Such a study should be made by a group combining legal, scientific, and business points of view, which can approach the problem judicially and without prejudice, and with ample time for its full consideration. The nature of the problem is brought out by the following:

There has been enormous change in technique and commercial practice in the last hundred years. The patent system at its inception contemplated an individual inventor, given a monopoly for 17 years as a reward and stimulant for invention, and to enable funds to be obtained for commercialization. This simple stipulation no longer obtains. What was originally a self-sufficient patent to an individual for 17 years has developed into a patent structure or assemblage of patents, giving a substantially permanent monopoly in an advancing art to an industry or a group of industries. The justification for the extension in a democratic country of an absolute monopoly to an inventor for 17 years, on the basis that this is a reasonable reward for his disclosure of his invention in lieu of maintaining it secret, no longer applies generally. In these days of intensified research and development it is the usual experience to find that important advances arise nearly simultaneously at many points. They are the result of an advancing knowledge and technique, and the advent of a specific human need and commercial opportunity. The individual inventor plays an important part in recognizing the situation and supplying the needed combination. In most cases, however, he could not hold it secret and use it privately if he wished. Moreover, if he did not appear with his invention it would not be long in these intense times before some other inventor would supply the necessary creative thought. This is not exclusively the situation, of course. There are still brilliant and striking flashes of intellect which create startling inventions which would not otherwise be made for perhaps a generation. The point is that inventions of this type are few and far between, and they are insignificant in number compared to the nearly 100,000 patents now issued annually. Moreover, most of these brilliant advances would be made and disclosed whether or not there were a patent system designed to produce a reward. The old justification for the extension of exclusive monopoly no longer holds.

There is still, however, a fully valid reason for continuing the system of extending a patent monopoly. New developments are hazardous. Only a small fraction of the attempts to bring into public use new and untried combinations are commercially successful. It is imperative that there should be an opportunity for the successful venture to reap a speculative profit. If it were assured only of a competitive profit, funds would not flow into new ventures, and this country would soon lose its place in a rapidly advancing technique. The opportunity for the necessary speculative profit can be secured only by the extension of a monopoly. Moreover there is great danger that an ill-advised restriction of this monopoly would cut the heart out of a system on which a great part of the striking industrial development of this country has been based.

Certainly a system of compulsory licensing based merely on failure to manufacture under a patent, such as has been in effect with dubious results in several countries, is not an adequate solution of the problem. A group which succeeds in arriving simultaneously at two new ways of adequately supplying a public need should not be penalized by being forced to manufacture both resulting devices.

Much of the difficulty arises because, under the law, all inventions are treated on an equal basis. A new collar button and a new flying machine result in patents granting similar rights and privileges. Careful consideration should be given to the desirability of creating two classes of patents, major and minor, with a relatively limited grant under the latter. A part of the distinction should result from the fact that some inventions are of such nature that they demand large and perilous expenditures, such as become expedient under monopoly, in order to bring them to fruition for the public benefit; whereas other inventions would come into use whether there were a patent system or not.

Under the present system, when a suit for infringement is successful, the court has no alternative than to assess profits and damages and order the cessation of infringement. When a patent has thus been found valid and infringed, the court cannot consider the public interest when called upon to issue an injunction to stop the use of the combination by others than the owner and his licensees.

Often the infringed patent is incidental or minor, or its primary utility may lie in an entirely different field. It would appear reasonable that in such cases, and to prevent unwarranted disruption of industry, the Court should be enabled to order the payment of reasonable royalties, rather than simply to order cessation. Such a provision would resolve the quandary in which Courts are forced by the strict letter of the law to act in a manner contrary to what appears to be broad public interest. Yet the determination that such a situation really existed would be difficult, and the evaluation of the extent to which a given patent controlled a given situation would be bound to be vague. In order to be definite such a change in our basic patent law as is here envisaged should therefore wait until the classification of patents into major and minor groups has been established, or until some equally positive way has been developed of delimiting the discretionary power of the courts.

The situation is thus a complicated one, in which hastily considered changes are highly inadvisable. It is believed that the modifications recommended in this report will result in a firmer base from which to approach the whole question of compulsory licensing.

SECONDARY MODIFICATIONS

The three major modifications recommended above are of primary importance. However, there are many secondary modifications, some of which are already receiving effective attention on the part of the Advisory Committee to the Patent Office, on which comment is in order.

Patent Office Personnel and Facilities: Every effort should be made to increase the standing and ability of the personnel of the Patent Office. They are handling an exceedingly difficult piece of work, which is an essential undertaking for the good of the country. In this work they should be generously supported. There are various ways in which they can be assisted, outside of the simple matter of remuneration. It appears desirable that examiners should have an opportunity to become acquainted with the developments in their field, by visits to industry and by further study, in order that they may perfect themselves in the handling of their advancing arts. They should have better library facilities. It appears also desirable that there should be a mechanism by which they may consult experts on scientific or technical questions, of course without disclosure of any matter under their consideration. They represent the public in important negotiations and the dignity of their position should be enhanced, and real accomplishment in this important public service recognized. It should receive direct subsidy in addition to all income from fees. The benefit to industry will return this investment tenfold.

Delays: The matter of delays is always serious. The burden which this places on industry at large is not always comprehended. Technical matters move much more rapidly in these days than they did a generation ago, and there is no inherent reason why legal matters should not also become accelerated. We are in a vastly different age from that when it took months to communicate with Washington. It would appear that the time allowed for the answer to an Office action and the time allowed before the payment of a final fee might with propriety be still further reduced. Similarly, the allowable delays in interferences should be cut down. Since, in American practice, the monopoly runs from the date of issue rather than the date of application, and since attorneys often delay the prosecution of applications in order thus to extend the effective monopoly, the Patent Commissioner should in the public interest rigorously restrict the pendency of applications and the duration of interferences to the minimum period consistent with proper examination and adjudication, and the Office rules should be modified wherever necessary to bring this about. These matters are receiving attention by the Advisory Committee to the Patent Office, together with others affecting the procedure in that

Office. Progress has been made, particularly in regard to interferences, and further progress is desirable.

Another type of delay occurs in connection with litigation. Wherever delays are unnecessary they should be studiously avoided, as they constitute a serious drag on industrial progress. There is a delay which sometimes occurs by reason of failure of a judge to give his decision promptly after the conclusion of a suit. It is realized that an interval at this time is necessary in order that a judge may read the law. However it appears that the interval which occurs between the conclusion of the suit and the rendering of the decision is often much longer than is necessary for this purpose. This appears to be often due to the difficulty experienced by the judge in fully understanding the technical facts presented to him, and in such cases the modification in court procedure recommended in this report will remove much of this difficulty. If delay occurs by reason of undue burden on the judge, then the burden on the court should be relieved in order that it may be reduced. It is entirely possible that some judges do not realize the serious harm which may be occasioned by delay, and that a better realization of this fact would automatically result in greater expedition. It is essential that delays be reduced at all points as far as is consistent with proper deliberate procedure, for the correction of the existing situation lies within the purview of the courts.

Joint Inventions: There is confusion regarding the matter of joint inventions. This is sometimes the reason why a patent becomes invalid on what is substantially a technicality. If the law states that the actual inventor must sign the application, but that he may be joined if he wishes by others who have in his opinion contributed, without danger of his patent's being found invalid because of the fact that their contribution is later found not to have been essential, the situation will be thoroughly clarified. This has been suggested many times.

Reissues, Disclaimers, Renewals: There seems to be strong argument for abolishing the granting of reissues and for simplifying the law concerning disclaimers. Expedition and clarity would also result if the practice of allowing renewals were discontinued. These matters appear to be in the nature of unnecessary complications, which confer a proper benefit in relatively rare instances, but the continuance of which in their present forms causes more confusion and cost to the public than is warranted by the results.

Equitable Treatment of American and Foreign Inventors: The American inventor is at a disadvantage in certain respects as compared to the foreign inventor. This whole situation is involved with the international agreements regarding patents. It requires careful study in order that any modifications introduced shall not give justifiable offense. However, the rights of the American inventor should be maintained on the same plane as of those in foreign countries who apply for United States patents, or for patents in other countries.

Reclassification of Patents: There has long been need for a thorough reclassification of patents in the Patent Office. The funds necessary for this piece of necessary work are not large, and should be provided.

Annual Taxes: There is much confusion due to the enormous number of patents in this country. As far as concerns those which are issued, not expired, considered valuable by their owners and yet probably invalid, little can be done except to leave the matter to litigation. There are, however, many issued patents which are now known to be worthless by those who hold them. It would be of great help if these could be removed from consideration. There are in various countries systems whereby patents are subjected to an annual tax. The result of this is promptly to remove from consideration all patents which are regarded by their owners to be not worth the payment of a tax. Such a system would greatly clarify the atmosphere in which industry operates by removing dead material, as patents upon which the tax remained unpaid would lapse.

The introduction of such a system should, however, be made in such a way as not to increase the burden of the individual inventor. In fact he is already overburdened financially by the present filing and final fees, taken together with his attorney's fees. The part played by individual and isolated inventors in our industrial development is not proportionately as great as it once was, for the greater part of modern invention comes from the joint work of many in laboratories. However, the day of the individual inventor is not past; his services to the country are needed and should be encouraged. He often points out the new and useful combination which would otherwise be overlooked if it were not for his discerning eye. The careers of successful individual inventors show that an expense at the time of making an invention is often a serious burden indeed, especially on the first invention of a series. Expense at a later stage, however, is not likely to be serious at all; for if the invention is truly important it soon attracts funds for its development, and further patent expenses in the Patent Office are a minor matter compared to the expense of such development, which is usually carried by others. It is desirable therefore that the initial burden on the inventor be reduced by cutting the filing and final fees, and that this be offset by imposing annual taxes.

There should then be a system of annual taxes, beginning several years after issuance, and on an ascending scale. These should be so adjusted that the total income from fees and taxes will be approximately the same as at present. By this means the burden will be no greater than at present, it will be placed where it can most readily be borne, and there will result the removal from consideration of a vast number of patents which are now simply an impediment.

STIMULATING NEW INDUSTRIES

The United States has developed marvelously in a technical way. Much of this advance has been due to the innate ingenuity of its people, and the patent system has been one of the main rocks on which the prosperity of the country has been erected. The character of the people has not changed, but the times have changed decidedly. Other countries, not previously technically minded, are going forward rapidly in this direction. Competition in technical affairs will be keen, and any nation which does not rapidly progress will drop into a secondary position in a technical world.   The patent system, built exceedingly wisely in the early days of our history, and developed carefully and skillfully in the hands of the Patent Office, the attorneys and the courts, is no longer completely in tune with modern conditions.   It should be maintained and strengthened.   Alteration is now essential if it is to continue to be a firm foundation for industrial advance.   Modifications should be entered upon carefully and thoughtfully, without destroying any vital part of the structure, but nevertheless courageously and thoroughly. Such a procedure is essential for the welfare of the country.

The greater problem of the stimulation of new industries is related to the more comprehensive program.   The benefits to be derived from a modernization of the patent system are dependent upon the treatment accorded by the people of this country to their industries generally.

The frontiers have disappeared.   No longer may a citizen break new ground beyond the horizon. But the opportunity for pioneering in the application of science to human needs remains, and calls for the same virtues of courage, independence, and perseverance.   It still is possible to enter uncharted regions in industry, and it is still hazardous thus to open new territory for the national welfare.

There has been a powerful trend toward stronger government control of large industry in recent years.   Unfortunately this has resulted in many measures which have borne heavily, and which have added artificial hazards to those naturally in the path of new ventures.   Independence has been curtailed. Legal complexities have been multiplied.   The making of a large profit has been frowned upon.   The creation of truly new industries and products has been rendered nearly impossible. This trend must reverse.

The removal of unnecessary hurdles in the patent system will help.   It can provide, however, only part of the essential correction. He who brings a new product or a new industry into being, with consequent gainful employment and a quickening of the national tempo, must be truly encouraged. As he takes great risks, and as many failures in new ventures are inevitable for each success, so must he feel secure in the earning of that speculative profit which is his incentive. It is the function of government to protect him from badgering by any organized group, so long as he regards the primary rights of others in his attempt to advance. Above all it is the function of government to see that he is constrained in his activities within the path of legitimate effort in as simple a manner as possible.

The patent system requires modification in this regard. But the welfare of the pioneer should be always prominently in mind wherever government control of industry is considered, in regulations concerning fair competition, in systems of taxation, in rules regarding the issuance of securities, and in all other control which affects him. Upon his progress depends the standing of our country in a shrunken world of intense competition, and the standard of living of our people compared to those of other lands. We sadly need to return to the realization that the pioneer is a benefactor, against whom the door of opportunity must not be closed.

16: SCIENCE FOR WORLD SERVICE

Thoroughout the world, thoughtful people are agreed that there must be peace, and that the nations must join together to maintain it. The bombs that burst over Hiroshima and Nagasaki require that we end war as an accepted and possible instrument for nations. In those bombs which unleashed the titanic power of atomic energy, mankind demonstrated to itself that it has reached the ultimate in destructive power. This means that a new world, evil or good, as we choose, lies before us. On the one hand are utter ruin and suicide; on the other are friendship and abundant life. It makes no difference whether a man is a physicist or a farmer, a Russian or an American, a youngster or a grandfather, a machinist or a financier—when he looks at these alternatives, he thinks first of all as a simple human being. And as he sums up in his mind the toll of death and devastation which the war has exacted, and projects in his imagination the horribly greater toll which another war would surely demand, he knows in his heart that greater than all other considerations—of race, nationality, ambition, trade, profession, power, or prestige—is the one all-encompassing fact that there must be peace and good will among men.

This has been mankind's problem after every war in history. It has never been solved—but there has always been another chance. This time it *must* be solved, for there will be no chance to try again. Another war would not necessarily wipe the human race from the face of the earth. But it could reduce the human race to a savage level or below.

The atomic bomb means that war now could come with volcanic suddenness and volcanic destructiveness to the headquarters of industry and production, and could blast the nerve centers of civilization into impotence even before an alert could be spread. The entire pattern in which we are accustomed to think of war is scrapped by this truth. Moreover, non-atomic weapons which were under study, in preparation, or in existence when atomic bombs ended this war could by themselves when fully developed obliterate civilization. The atomic bomb emphasized and punctuated a stark reality. It is therefore imperative for us finally to prove that the old assumption that wars are inevitable is a fallacy.

Already we have made marked progress toward that proof, in the establishment of a framework within which the peaceful people of the world can work together. The United Nations Organization is a fact, and a fact that can be made as powerful in its way as the fact of the availability of atomic energy.

I believe that a strong United States is the surest guarantee of peace in the world. Not a strong United States policing the world. Not a United States strong in merely immediate military power designed to wage war in the next few years. I mean a United States strong now, and stronger in the future, governmentally, economically, scientifically, industrially, morally, and also, and until we arrive at an international framework capable of preserving peace, strong in a military sense.

We must grasp the tough fact that the very emphasis on peace in the great democracies in the interval between the last two wars undoubtedly fostered the aggressors' conviction that the democracies were soft and decadent, and encouraged Hitler to strike. Talk of peace must this time be realistic; we shall need to maintain our full strength as a military power if we are to be respected and listened to.

Americans who sincerely want to share in securing and maintaining the peace of the world through a strong international organization have their work plainly before them. We must all judge our courses as good or bad according to how they will help or hinder the strength with which the nation goes forward, not as a possible combatant in a world of isolated suspicious states, but as one free people among other free peoples. This applies to the engineer in the field, the scientist in the laboratory, the industrialist at his desk, the mechanic at the lathe—to all of us as individuals, and to all of us as groups. No individual, no group, has any right to override the good of the nation as a whole. At this critical time any individual who places a selfish interest above the good of his country, above its strength in a difficult world, should pause to think that by so doing he may be making it necessary for his children or grandchildren to fight in a desperate war.

The realistic sense of moral responsibility which I stress here is basic. With it, we have the foundation on which the bulwarks of national strength can be built.

I place high in the list of these the maintenance of vigorous research in fundamental science, with Federal financial aid for the support of research programs and for the education of future scientists, and with stress on the fact that fundamental research demands at the same time the highest degree of freedom and initiative for the individual. There must be no taint of regimentation as Federal support is thus provided. The Congress now has under consideration bills which will establish a national research foundation. Through this mechanism, the people of the United States will be enabled to foster the fundamental studies of science from which come rich basic knowledge and immediate practical utility. Of particularly great importance is the fact that this legislation will help give the brains of young Americans a full chance to work. Throughout all the fields of knowledge, and especially in the vast unknown new field of atomic energy, the need and the opportunity for young fresh minds are greater than ever before in history.

During the war years we drew heavily on our scientific capital, making great advances in applied science—in radar, rockets, antiaircraft gunnery, in immediate therapeutics, in transportation. To do so, we had to give up fundamental research and so we had to sacrifice the future to the present. We must now replenish the reservoirs of fundamental knowledge. It is through the application of the results of vigorous fundamental research that we have in the past created extensive industries, secured productive employment for our people, raised our standard of living and of general education, and increased the national income upon which government draws for the general good. We must be able to rely in the future on fundamental science to provide the basis for these things to a greater extent than we have in the past.

The advantages which an alert and aggressive military establishment gained from scientific research in the hard-fought struggle just past need no rehearsal here by me. Bear in mind the third member of the powerful team that made them possible. This third member is American industry, flexible, resourceful, vigorous. Without it, this country and the rest of the free world might well have gone down in defeat. Industrial might, when well exerted through teamwork between management and labor, has always been one of this country's greatest sources of national strength, in peace or war. Hence I hold that we must foster and preserve the industrial zeal that has served us well ever since our pioneer days. Only thus can we be sure of the avenues by which the results of scientific and engineering progress find their place as useful products in national and international life.

This does not mean that the central government should relax its vigilance to protect the public against wrongdoing—in business or elsewhere. It does mean, however, that we must encourage the advent of new industrial units, remove obstacles and petty annoyances, and thus create a climate of opinion in which sound business will thrive. It does not mean failure to regulate where regulation is necessary, as in the natural monopolies, but it does mean that there should be no hostility between government and business, even if business is big.

We must have a strong government. Its several functions must be clearly defined and allocated, it must consist of able, alert men, and above all it must be honest and upright because its members have moral integrity and a breadth of vision which lets them see against the background of history the significance of what they do. In the war years, all Americans in and out of uniform have been alert participants in their government. It will not do for that active sharing to slacken now.

In this war the representative republican system proved efficient, versatile, adaptable. We must keep it that way. There is no inconsistency between this assertion and the existence of a strong federal government in the United States. As society becomes more complex there are more things for the central government to do—many things which only it can do. Nor is there inconsistency between this assertion and the creation of great government works. Some of these are as old as the nation—consider the postal system, the highways spanning the continent. Such works as these are of necessity a national and hence a governmental responsibility. A new one is before us in the control and administration of atomic energy. Already, in legislation now being considered, we are undertaking the problem of national control—an essential first step toward ultimate world regulation and utilization of this vast new power. Establishment of an Atomic Energy Control Commission of competent and disinterested citizens armed by the Congress with the unprecedented powers demanded by their unprecedented responsibilities, with proper safeguards against their arbitrary or unreasonable use, should soon be expected.

Now just as the self-interest of groups within the nation must yield to the essential requirement that the nation be strong, so the self-interest of nations must in the years to come be subordinated in order that the world organization may be strong. We must be prepared to recognize that no short-range self-interest of the United States can be allowed to stand in the way of full and sincere collaboration with other nations in the furtherance of peace. At this moment, we Americans face this issue, and the issue is stated in very great terms. It is the question whether and how we shall share with other nations our knowledge of the control and utilization of atomic energy. I do not use the atomic bomb as a term in stating the issue; the bomb is but one application of atomic energy, a titanic and awesome application, it is true, but still only one application.

Let me say something about the bomb. First, we can never be thankful enough that the secret was learned by peace-loving peoples, not by the fascist nations which sought with all their might to master it in order to unleash atomic war on the whole world. Second, the bomb did not win the war; the bomb did end the war swiftly and I think mercifully, and thereby saved many thousands of human lives. Third, and this is extremely important, by reason of its sudden, spectacular effectiveness, the atomic bomb underscored and emphasized as never before the fact that the nations of the earth must put an end to all wars for ever. Fourth, and of equal import, the development of the bomb—the work of a congress of free minds and free hands in a free country—is undying testimony to the strength and vitality of the philosophy of government in which we believe. Only in a free country where people have faith in the good will of one another could so vast an undertaking have been carried through so successfully in so short a time. Whatever I have said earlier about the teamwork of science, management, labor, the military, and government is forcefully re-emphasized by this achievement.

For the time being—that is, for the five to fifteen years which other nations must have if they are to make atomic bombs of their own—the control of atomic energy thus achieved by scientists and engineers working in the United States puts in the hands of the American people a power and an opportunity such as no people ever before have held. It is a startling fact, but it is a fact, that the United States today, if it would continue to put all its resources into such an effort, and if matters of physical force only were involved, could turn aggressor, devastate the centers of the world with atomic bombs, and in a short time impose its will on all nations. The power is thus the power to rule the world. The opportunity is the opportunity not to use that power for that purpose. These are not alternatives for a free people rich in the tradition of freedom.

For us, there is no choice between them. The United States will use the opportunity, not the power.

It is perfectly plain that we will not use atomic energy to impose our will physically on the world. How shall we use it? Remember that though all we have now is a bomb, control of atomic energy will in time become an economic factor of the first importance. Note that in mankind's history the applications of chemical energy appeared first at the simplest level—uncontrolled fire, then explosions—and that later they were one by one controlled and applied to meeting human needs. The science of atomic energy today is comparable to the science of electricity in the time of Faraday. Things move faster today than they did a century ago; hence we may expect that though it is far more difficult, the development of atomic energy for peaceful industrial and economic use will be swifter than was the development of electricity. The atom should be at useful constructive work for us within ten years. Yet it will not perceptibly alter the pattern of our living until long after that. We shall still use steam and electricity for lighting and power. The atom will generate these things—not drive the family car. And we must bear in mind, as we look thus to the future, that this development which the public interest requires will have to be carried on under controls which the public safety demands. Using atomic energy is a dangerous undertaking. But that energy must be brought to use, for it can ultimately bring all mankind more ease and peace than we have ever known. American industrial ability, working within sensible limits which the well-being of the people imposes, can be relied on to overcome the dangers and produce the benefits.

As an engineer, I have good reason to know that the free exchange of ideas and knowledge is the first requirement of progress. We had such a system before the war, and under it American science and engineering led the world. It must be restored, for only by the cross-fertilization of brains do we breed great thinking, and the peaceful control of the energy of the atom will demand much great thinking. Therefore, I hope to see the United States make the first great move toward the renewal of international exchange of scientific knowledge. I believe we should undertake to share with our world partners all of our basic scientific knowledge of atomic energy.

This does not mean that I think we should blandly give away this part of the information without consideration. The consideration we should ask is quite clear: We would expect all nations to do likewise, in fact to be equally candid and open in all areas of scientific progress, and to make the interchange real by constant exchange of scientists, students, and publications, with a policy of open doors in scientific laboratories all over the world. We can bring about this result if we are united and wise, but it should be clear that there are two great objects: first, to open our own scientific doors, and second, to do so in such manner that all other doors will also open.

This would be only a first step, but it would be an important one. It would not involve "giving away the secret of the atomic bomb." That resides, as far as it exists, primarily in industrial experience, the solving of a multitude of practical problems, and the intricate techniques of application. It cannot be too strongly emphasized that no man could convey this information by a formula or a diagram or two, if he would; it is much too complex for that. I would propose that we retain such information securely at present, for it is one of the reasons that our lead over other nations that might wish to construct atomic bombs is as large as it is. But the first step of suggesting full interchange on the basic science of the atom is still important; for to take it would indicate strongly that we wish to proceed down the road of international collaboration.

Other steps can follow, if the first is successful. I would advocate as the next step the placing of full information in regard to all aspects of atomic energy in the hands of a body in the United Nations Organization, with instructions for the complete dissemination of it. But I would do so only if there were complete acceptance of the principle that this body would have inspection rights, to be implemented by a scientific board internationally constituted, and before the door was opened wide I would make very sure that such inspection would work, so that no nation could in fact proceed in secret with military applications.

Beyond this lie other steps, and these hardly need to be specifically formulated as yet, for the road is long. Trusteeship for all military embodiments, provisions for use only under the orders of the Security Council, dispersion of materials into industrial embodiments from which they could not be reassembled without the knowledge of the world, are possibilities to explore.

The first step toward that ultimate goal is to establish the full free flow of facts—the complete and honest exchange of knowledge—among the nations. We Americans are the first to be in full possession of the most powerful and most precious knowledge of the physical world ever discovered. We can therefore open the way toward assuring the growth of the full free flow of facts, by providing for complete interchange of information on the basic scientific aspects of atomic energy. In the War Department release in early August, we made the right start, and I believe we should go farther on that course. The strongest help that we can give to bettering the common destiny of mankind is to demonstrate in this way our faith in the good will of men and our desire to be one nation among other nations in a peaceful world of free interchange.

As we enter a new world I urge two things. Let us maintain our country strong in every way. In this strength, let us lead through the path of international understanding to the organization of a sovereign world where the nations of men will find the better, happier life which man's final mastery of the energy of the atom offers, if he will but use it rightly. To this end every citizen, in every contact with his fellows, guided by his patriotism and his own judgment, has his duty clearly before him.

17: THE BUILDERS

The process by which the boundaries of knowledge are advanced, and the structure of organized science is built, is a complex process indeed. It corresponds fairly well with the exploitation of a difficult quarry for its building materials and the fitting of these into an edifice; but there are very significant differences. First, the material itself is exceedingly varied, hidden and overlaid with relatively worthless rubble, and the process of uncovering new facts and relationships has some of the attributes of prospecting and exploration rather than of mining or quarrying. Second, the whole effort is highly unorganized. There are no direct orders from architect or quarrymaster. Individuals and small bands proceed about their businesses unimpeded and uncontrolled, digging where they will, working over their material, and tucking it into place in the edifice.

Finally, the edifice itself has a remarkable property, for its form is predestined by the laws of logic and the nature of human reasoning. It is almost as though it had once existed, and its building blocks had then been scattered, hidden, and buried, each with its unique form retained so that it would fit only in its own peculiar position, and with the concomitant limitation that the blocks cannot be found or recognized until the building of the structure has progressed to the point where their position and form reveal themselves to the discerning eye of the talented worker in the quarry. Parts of the edifice are being used while construction proceeds, by reason of the applications of science, but other parts are merely admired for their beauty and symmetry, and their possible utility is not in question.

In these circumstances it is not at all strange that the workers sometimes proceed in erratic ways. There are those who are quite content, given a few tools, to dig away unearthing odd blocks, piling them up in the view of fellow workers, and apparently not caring whether they fit anywhere or not. Unfortunately there are also those who watch carefully until some industrious group digs out a particularly ornamental block; whereupon they fit it in place with much gusto, and bow to the crowd. Some groups do not dig at all, but spend all their time arguing as to the exact arrangement of a cornice or an abutment. Some spend all their days trying to pull down a block or two that a rival has put in place. Some, indeed, neither dig nor argue, but go along with the crowd, scratch here and there, and enjoy the scenery. Some sit by and give advice, and some just sit.

On the other hand there are those men of rare vision who can grasp well in advance just the block that is needed for rapid advance on a section of the edifice to be possible, who can tell by some subtle sense where it will be found, and who have an uncanny skill in cleaning away dross and bringing it surely into the light. These are the master workmen. For each of them there can well be many of lesser stature who chip and delve, industriously, but with little grasp of what it is all about, and who nevertheless make the great steps possible.

There are those who can give the structure meaning, who can trace its evolution from early times, and describe the glories that are to be, in ways that inspire those who work and those who enjoy. They bring the inspiration that not all is mere building of monotonous walls, and that there is architecture even though the architect is not seen to guide and order.

There are those who labor to make the utility of the structure real, to cause it to give shelter to the multitude that they may be better protected, and that they may derive health and well-being because of its presence.

And the edifice is not built by the quarrymen and the masons alone. There are those who bring them food during their labors, and cooling drink when the days are warm, who sing to them, and place flowers on the little walls that have grown with the years.

There are also the old men, whose days of vigorous building are done, whose eyes are too dim to see the details of the arch or the needed form of its keystone, but who have built a wall here and there, and lived long in the edifice; who have learned to love it and who have even grasped a suggestion of its ultimate meaning; and who sit in the shade and encourage the young men.

INDEX

HISTORY, PHILOSOPHY AND SOCIOLOGY OF SCIENCE

Classics, Staples and Precursors

An Arno Press Collection

Aliotta, [Antonio]. **The Idealistic Reaction Against Science.** 1914

Arago, [Dominique François Jean]. **Historical Eloge of James Watt.** 1839

Bavink, Bernhard. **The Natural Sciences.** 1932

Benjamin, Park. **A History of Electricity.** 1898

Bennett, Jesse Lee. **The Diffusion of Science.** 1942

[Bronfenbrenner], Ornstein, Martha. **The Role of Scientific Societies in the Seventeenth Century.** 1928

Bush, Vannevar. **Endless Horizons.** 1946

Campanella, Thomas. **The Defense of Galileo.** 1937

Carmichael, R. D. **The Logic of Discovery.** 1930

Caullery, Maurice. **French Science and its Principal Discoveries Since the Seventeenth Century.** [1934]

Caullery, Maurice. **Universities and Scientific Life in the United States.** 1922

Debates on the Decline of Science. 1975

de Beer, G. R. **Sir Hans Sloane and the British Museum.** 1953

Dissertations on the Progress of Knowledge. [1824].
2 vols. in one

Euler, [Leonard]. **Letters of Euler.** 1833. 2 vols. in one

Flint, Robert. **Philosophy as Scientia Scientiarum and a History of Classifications of the Sciences.** 1904

Forke, Alfred. **The World-Conception of the Chinese.** 1925

Frank, Philipp. **Modern Science and its Philosophy.** 1949

The Freedom of Science. 1975

George, William H. **The Scientist in Action.** 1936

Goodfield, G. J. **The Growth of Scientific Physiology.** 1960

Graves, Robert Perceval. **Life of Sir William Rowan Hamilton.**
3 vols. 1882

Haldane, J. B. S. **Science and Everyday Life.** 1940

Hall, Daniel, et al. **The Frustration of Science.** 1935

Halley, Edmond. **Correspondence and Papers of Edmond Halley.** 1932

Jones, Bence. **The Royal Institution.** 1871

Kaplan, Norman. **Science and Society.** 1965

Levy, H. **The Universe of Science.** 1933

Marchant, James. **Alfred Russel Wallace.** 1916

McKie, Douglas and Niels H. de V. Heathcote. **The Discovery of Specific and Latent Heats.** 1935

Montagu, M. F. Ashley. **Studies and Essays in the History of Science and Learning.** [1944]

Morgan, John. **A Discourse Upon the Institution of Medical Schools in America.** 1765

Mottelay, Paul Fleury. **Bibliographical History of Electricity and Magnetism Chronologically Arranged.** 1922

Muir, M. M. Pattison. **A History of Chemical Theories and Laws.** 1907

National Council of American-Soviet Friendship. **Science in Soviet Russia: Papers Presented at Congress of American-Soviet Friendship.** 1944

Needham, Joseph. **A History of Embryology.** 1959

Needham, Joseph and Walter Pagel. **Background to Modern Science.** 1940

Osborn, Henry Fairfield. **From the Greeks to Darwin.** 1929

Partington, J[ames] R[iddick]. **Origins and Development of Applied Chemistry.** 1935

Polanyi, M[ichael]. **The Contempt of Freedom.** 1940

Priestley, Joseph. **Disquisitions Relating to Matter and Spirit.** 1777

Ray, John. **The Correspondence of John Ray.** 1848

Richet, Charles. **The Natural History of a Savant.** 1927

Schuster, Arthur. **The Progress of Physics During 33 Years (1875-1908).** 1911

Science, Internationalism and War. 1975

Selye, Hans. **From Dream to Discovery: On Being a Scientist.** 1964

Singer, Charles. **Studies in the History and Method of Science.** 1917/1921. 2 vols. in one

Smith, Edward. **The Life of Sir Joseph Banks.** 1911

Snow, A. J. **Matter and Gravity in Newton's Physical Philosophy.** 1926

Somerville, Mary. **On the Connexion of the Physical Sciences.** 1846

Thomson, J. J. **Recollections and Reflections.** 1936

Thomson, Thomas. **The History of Chemistry.** 1830/31

Underwood, E. Ashworth. **Science, Medicine and History.** 2 vols. 1953

Visher, Stephen Sargent. **Scientists Starred 1903-1943 in American Men of Science.** 1947

Von Humboldt, Alexander. **Views of Nature: Or Contemplations on the Sublime Phenomena of Creation.** 1850

Von Meyer, Ernst. **A History of Chemistry from Earliest Times to the Present Day.** 1891

Walker, Helen M. **Studies in the History of Statistical Method.** 1929

Watson, David Lindsay. **Scientists Are Human.** 1938

Weld, Charles Richard. **A History of the Royal Society.** 1848. 2 vols. in one

Wilson, George. **The Life of the Honorable Henry Cavendish.** 1851